RUSSIAN ENERGY IN A CHANGING WORLD

Russian Energy in a Changing World

What is the Outlook for the Hydrocarbons Superpower?

Edited by

JAKUB M. GODZIMIRSKI
Norwegian Institute of International Affairs (NUPI), Norway

LONDON AND NEW YORK

First published 2013 by Ashgate Publishing

Published 2016 by Routledge
2 Park Square, Milton Park, Abingdon, Oxfordshire OX14 4RN
711 Third Avenue, New York, NY 10017, USA

First issued in paperback 2016

Routledge is an imprint of the Taylor & Francis Group, an informa business

British Library Cataloguing in Publication Data
A catalogue record for this book is available from the British Library

The Library of Congress has cataloged the printed edition as follows:
Godzimirski, Jakub M.
Russian energy in a changing world : what is the outlook for the hydrocarbons superpower? / by Jakub M. Godzimirski.
   pages cm
 Includes bibliographical references and index.
 ISBN 978-1-4094-7028-1 (hardback)
 1. Energy policy–Russia (Federation) 2. Energy industries–Russia
(Federation) 3. Petroleum industry and trade–Russia (Federation) I. Title.
 HD9502.R82G63 2013
 333.790947–dc23

                                                            2013019357

ISBN 13: 978-1-138-27978-0 (pbk)
ISBN 13: 978-1-4094-7028-1 (hbk)

Contents

List of Figures

List of Tables

List of Contributors

Dr Derek Averre is Senior Lecturer and was between 2008 and 2012 Director of CREES (The Centre for Russian and East European Studies), University of Birmingham. CREES is one of the world's leading research institutes in the area of Russian studies and Dr Averre has been publishing extensively on the Russia related issues, specializing in Russian foreign and security policy, Russia–Europe relations and arms control and non-proliferation issues in the USSR successor states. He has published a co-edited book *New Security Challenges in Postcommunist Europe: Securing Europe's East* and numerous journal articles on these themes in International Affairs, European Security, Problems of Post-Communism, Europe-Asia Studies and Demokratizatsiya, as well as several book chapters. He has presented widely at both academic and policy-related conferences and seminars in the UK (including Wilton Park and Chatham House), Europe and North America.

Dr Pavel Baev is Research Professor at PRIO (Peace Research Institute in Oslo). He graduated from Moscow State University in 1979 and after having served at Research Institute of the USSR Ministry of Defence and Institute of Europe of the Russian Academy of Sciences, he moved to Oslo where he since 1992 has had various positions at PRIO. Dr Baev is a prolific writer publishing extensively both in the West and in Russia. Among his recent publications the most relevant for this project is his book *Russian Energy Policy and Military Power – Putin's Quest for Greatness* published in 2008 with Routledge.

Dr Irina Busygina is Professor at the Department of Political Science, Moscow State Institute of International Relations (University) – MGIMO (U) and the Director of Center for Regional Political Studies MGIMO. Her research interests include comparative politics, comparative federalism, globalization and regionalism, European integration and German politics. She graduated from the Faculty of Geography at the Moscow State University and followed her academic career at the Institute of Europe, Russian Academy of Sciences and at MGIMO. She has published extensively on Russian federalism regionalism and Russia's relations with Europe, both in Russia and in the West.

Dr Mikhail Filippov is Associate Professor of Political Science at the Binghamton University (NY). Professor Filippov holds an M.A. in Political Science from the University of California at Riverside, an M.S. in Economics and Political Science from the California Institute of Technology, and a Ph.D. in Economics and Political Science from the California Institute of Technology. His work focuses on comparative federalism, intergovernmental relations, and European politics and on contractual aspects of federal arrangements, selection and implementation of jurisdictional delineation in democratic federations, and the role of political agency in federal survival. His book *Designing Federalism: A Theory of Self-Sustainable Federal Institutions*, co-authored with Peter Ordeshook and Olga Shvetsova, was published by Cambridge University Press and received an Honorable Mention for the William H. Riker Prize of the Political Economy Section of the American Political Science Association in 2005.

Daniel Fjærtoft is a Russia expert combining intensive in-field experience and analytical skills. Fjaertoft holds a Master's Degree in the Philosophy of Economics from the University in Oslo where he wrote an empiric assessment of the micro-economic foundations of Norwegian-Russian trade. Fjaertoft has previously worked as an adviser with the Norwegian Barents Secretariat promoting Russian-Norwegian cooperation within the fields of business, petroleum and maritime infrastructure and transport.

Dr Jakub M. Godzimirski holds a PhD in social anthropology and is Research Professor with the Department for Russian and Eurasian Studies at the Norwegian Institute of International Affairs (NUPI www.nupi.no). In 2009–2010 he was also the head of the NUPI Energy Programme. His present research interests include Russian foreign and security policy, with focus on energy's role and Russia's relations with the West. He has edited several volumes on those issues and published many peer reviewed articles and book chapters on Russian energy policy and its reception in Europe.

Professor Valeriy Kryukov is Head of Department of Power and Commodity Markets at the Higher School of Economics, Moscow, Faculty of World Economy and International Affairs. Professor Kryukov has Doctor of Science degree from Institute of Economics and Industrial Engineering of the Siberian Branch of the USSR Academy of Sciences. Over the last years Professor Krykov has been publishing extensively on the development of the Russian energy sector.

Ms Eini Laaksonen is Master of Economic Sciences from Turku School of Economics and works as Project Researcher at the Pan-European Institute in Turku School of Economics at the University of Turku in Finland. She has been working in the Institute since 2008 and has been involved in various research projects. She has specialised in international business, the main areas of research interest comprising the future of Russian energy production, business opportunities in the Russian North, and the political risks of Russian energy industry for foreign investors.

Lars Petter Lunden holds MA in economics and finance from Universidade de São Paulo (Brasil), MA in economics from University in Oslo (UiO) and studied also at Boston University – School of Management. After completing his formal education Lunden worked at A&D Analyst at Statoil and as consultant at Econ Pöyry where his main area of expertise was Russian energy sector and international energy markets. Since 2011 he is – together with Fjaertoft – partner at Sigra Group, a Norwegian business development company focused on creating sustainable growth opportunities for Norwegian and Russian partners.

Mr. Kari Liuhto is Professor in International Business (specialisation Russia), Director of the Pan-European Institute at the University of Turku, Finland, and Director of Finland's Baltic Sea region think tank called Centrum Balticum. Liuhto's research interests include EU-Russia economic relations, energy relations in particular, foreign investments into Russia and the investments of Russian firms abroad, and Russia's economic policy measures of strategic significance. Liuhto has been involved in several Russia-related projects funded by Finnish institutions and foreign ones, such as the Prime Minister's Office, various Finnish ministries and the Parliament of Finland, the European Commission, the European Parliament, and the United Nations.

Dr Tatyana Mitrova is the Head of the Center for International Energy Markets Studies, Energy Research Institute (ERI) of the Russian Academy of Sciences and is currently an Assistant Professor at the Higher School of Economics, Moscow and Gubkin Oil and Gas University, Moscow as well as the head of the Global Energy department of the Skolkovo Energy Center (SEneC) at the prestigious Skolkovo Moscow School of Management. Dr Mitrova's primary responsibility is analyzing the development of global energy markets and the Russian Federation's energy export and import policy. She has been working closely with Russian ministries, as well as major Russian and international energy companies (Gazprom, RAO UES, TNK-BP, NOVATEK, Statoil, GDF SUEZ, Wintershall).

Arild Moe holds a cand.polit degree from the University of Oslo (political science, Russian, public law) and serves as Deputy Director/Senior Research Fellow at the Fridtjof Nansen Institute (FNI) in Oslo. His main research interests are Russian oil and gas industry; the regional dimension in the Russian petroleum sector; offshore activities in the Barents Sea; Russian climate politics, oil companies and CSR and Norwegian policy in the High North.

Ms Hanna Mäkinen holds Master of Arts in General History, Political Science and Contemporary History from the University of Turku and currently works as Project Researcher at the Pan-European Institute, Turku School of Economics at the University of Turku in Finland. She has been working in various research-related positions at the Pan-European Institute since 2008. Her main research interests include EU-Russia energy relations, in particular concerning natural gas, and politics and contemporary history of the Baltic States.

Acknowledgements

Jakub M. Godzimirski

The editor and authors of this book would like to thank the Research Council of Norway (RCN) for generous funding provided to the project Russian and Caspian Energy Developments RUSSCASP. The project, realized between 2008 and 2012 and coordinated by Arild Moe, has gathered a group of energy experts from the Fridtjof Nansen Institute (FNI), the Norwegian Institute of International Affairs (NUPI), ECON, the Peace Research Institute in Oslo (PRIO), the Bodø Graduate School of Business and the Higher School of Economics, Moscow who have worked closely together with a network of Russian and Western experts in the field. The project had three key foci, looking at Russia and the Caspian region as arenas for foreign energy companies, trying to understand the main driving forces and conditions for Russian and Caspian energy exports and looking at energy developments in the Russian High North. The texts presented in this book are a result of this international cooperation that was made possible thanks to the funding provided by the Research Council of Norway.

This book would not be possible without a successful cooperation between the Norwegian project team and its international partners who deserve our special gratitude not only for their contributions, some of which are presented in this volume, but also for sharing with us their deep insights and presenting a more nuanced picture of the Russian energy reality. We hope very much that this fruitful cooperation will continue in the many years to come.

The editor of this book would also like to express his special gratitude to his closest colleagues at NUPI who have given him support in moments of despair and have shown patience when things did not develop as they should. Especially Helge Blakkisrud, Indra Øverland and Elana Wilson Rowe deserve these warm words of appreciation, but also other NUPI friends have made me enjoy my time at NUPI's sixth floor in this hectic period.

It is impossible to forget about the contribution that has been crucial for this book becoming a reality – namely the tremendous work done on the manuscript by our language editor Susan Høivik. She has not only polished our English but has also – hopefully – made reading this book a smoother experience. Also our NUPI interns, and especially Shemshat Kasimova, provided me with practical help and made my work on this book less stressful.

However, production of this book was a rather stressful experience to my closest circle – my wife Magdalena, our son Jan and our daughter Zuzanna met this challenge with a lot of understanding and I appreciate their positive attitude.

The book is a result of a collective effort of the team of authors who have been working on that project for a long time. In the last phase of our work we also received valuable editorial help from the Ashgate team – Rob Sorsby and Margaret Younger – who have guided us through this editorial process. Although each of the authors is responsible for his or her contribution, the overall responsibility for that volume rests on my shoulders and any of its editorial shortcomings should be blamed on the editor alone.

Jakub M. Godzimirski, Oslo, December 2013

List of Abbreviations

AFA	Address to Federal Assembly
APR	Asia–Pacific Region
BPS	Baltic Pipeline System
Bcm	billion cubic meters
BTC	Baku-Tbilisi-Ceyhan
CEO	Chief Executive Officer
CIS	the Commonwealth of Independent States
CPC	Caspian Pipeline Consortium
EEZ	Exclusive Economic Zone
EIA	Energy Information Agency
ESPO/VSTO	East Siberia Pacific Ocean Pipeline
EU	European Union
FNI	Fridtjof Nansen Institute
FTS	Federal Tariff Service
GDP	Gross Domestic Product
IEA	International Energy Agency
IISS	International Institute of Strategic Studies
INSOR	Institute of Contemporary Development
JCC	Japanese Crude Cocktail
JV	Joint venture
LNG	Liquified Natural Gas
MET	Mineral Extraction Tax
MPR	Ministry of Natural Resources and Ecology
NATO	North Atlantic Treaty Organisation
NDPI	Mineral Extraction Tax
NG	Nezavisimaya Gazeta
NPT	Nadym-Pur-Taz

NUPI	Norwegian Institute of International Affairs
PfM	Partnership for Modernization
PRIO	Peace Research Institute in Oslo
RAS	Russian Academy of Sciences
RCN	Research Council of Norway
RIA	Russian Information Agency
RUSSCASP	Russian and Caspian Energy Developments Project
SCO	Shanghai Cooperation Organization
Tcm	thousand cubic meters
TsKR	Central Commission for Development of Mineral Fields
UK	United Kingdom
USA	United States of America
USD	US dollars
WTO	World Trade Organisation

Introduction:
Understanding Russian Energy after the Crisis

Jakub M. Godzimirski

The Russian Federation is truly an energy giant. Its known reserves of oil stand at 12.1 billion tons, or 5.3 per cent of total world reserves; in gas it controls an estimated 44.6 trillion m³ (21.4 per cent of known global gas reserves); and it also has an estimated 157 billion tons of coal, or 18.2 per cent of the world's known reserves. With those known reserves and at current production levels, Russia's production of oil can continue for the next 23.5 years; gas production can be maintained at the same level for next 73.5 years, and coal production for the next 471 years (all figures from BP, 2012).

In 2011 Russia produced 334 million tons of coal, 511 million tons of oil, 669 billion cubic meters (bcm) of natural gas and 1052 billion kWh of electricity. Oil production was 0.8 per cent higher in 2011 than in 2010, and gas production increased by 2.9 per cent compared to the previous year, while production of coal and electricity grew by 2.8 and 1.4 per cent, respectively. In 2011, Russia exported 203 bcm of gas, 241.8 million tons of oil and 104.6 million tons of coal (Ministry of Energy, 2012). This means that Russia exported 30.3 per cent of the gas, 47.5 per cent of the oil and 31.3 per cent of the coal produced in the country.

In the record year 2008 the value of energy exports from Russia reached USD 307.4 billion – an impressive 42 per cent higher than the previous year. In 2008 energy resources accounted for 65.7 per cent of total exports from Russia (WTO, 2011), and the revenues generated from the production and sales of energy commodities represented almost 50 per cent of the country's budget revenues.

The European Union remains the most important export market for Russian energy commodities, receiving 88 per cent oil, 70 per cent gas and 50 per cent coal exported from Russia in 2010. Imports from Russia represented 34 per cent of gas imports to the EU, covering 23 per cent of gas consumption there. Also, 33 per cent of the crude oil imported into the EU came from Russia, covering 30 per cent of EU oil needs. In addition Russia supplied 23 per cent of the petroleum products and 30 per cent of the coal imported into the EU (figures from Ministry of Energy, 2011).

The abundance of energy commodities is both one of Russia's most important competitive advantages and one of its major challenges, creating many tensions and economic, social and political problems. Because, as Clifford Gaddy (2011) put it, 'oil and gas are something everyone wants, and Russia has more of them than anyone else', being so richly endowed with energy resources has a direct impact on the country's foreign and security policy: on the one hand, it is important to promote Russia's economic interests – but it is also essential to protect its strategic assets in a world where competition for access to and control over energy resources has become a key international policy driver (on those aspects see Orttung and Overland, 2011).

This book aims to provide a better understanding of the challenges that Russian decision makers have had to cope with in formulating and implementing energy policy in the post-crisis situation, and to present a *tour d'horizon* of issues that have risen to prominence in the wake of the economic crisis. In addition, the contributors explore how the crisis has changed the framework conditions in which Russian energy policy is being realized and how those framework conditions influence the country's energy policy and its energy relations with other actors. The chapters in this volume examine the importance of the energy sector in shaping Russia and its relations with the outside world in today's post-crisis situation.

Between 2007 and 2009 the tone in the Russian debate on energy policy changed dramatically. Prior to the crisis Russia had been called 'an indispensable energy power' (Hill, 2004) or even an energy superpower (IISS, 2006, Rutland, 2008) – but in 2009 President Medvedev described Russia's dependence on energy commodities as a 'humiliating addiction', and proceeded to launch an ambitious modernization programme aimed at lessening the role of the energy sector in the economy and making the national economy more diversified and innovative (Medvedev, 2009). This modernization programme had various functions – it was intended to help Russia become a more competitive actor, but it was also a political platform used by Medvedev and his team to strengthen their position on the Russian political map and to present an alternative to the plans of Vladimir Putin.

The economic crisis was indeed a painful and sobering experience for the Russian political elite. When the oil price reached almost USD 150 per barrel in the first week of July 2008 and Russia could cash in on enormous oil windfalls while the West started facing what Moscow saw as a deep crisis of the Western model of economic development, Kremlin leaders could hope for better times for their country. However, only one year later, during the summer months of 2009, Russia found itself hit by the deepest economic crisis in the Putin era. It was no longer talk about *slower* growth of the Russian economy – it had started to *shrink*.

On 24 June 2009 the Russian media informed about the economic results of the first months of 2009. This made for depressing reading. *Vedomosti* reported that the country's economy had shrunk by 10.2 per cent in the first five months of 2009, and the results for May alone were 11 per cent worse than for the same month in 2008. Deputy Economic Development Minister Andrei Klepach was forced to conclude that the overall decline was continuing, with no turnaround as yet. He also underlined that it would be very difficult to prevent a greater drop in GDP than the 6–8 per cent expected by the Ministry. According to various estimates, the Russian economy was at that stage expected to shrink between 6 per cent (IMF) and 7.5 per cent (World Bank/EBRD); some Russian officials even admitted that the economy could shrink by 9 per cent that year – which proved to be not far from the mark.

When the Russian State Statistics Committee finally presented the official data on economic activity in the first six months of 2009, it became clear that the crisis had continued and seemed deeper than originally estimated. In June 2009 the Russian economy shrank by 9.6 per cent compared with the same month in 2008; for the first six months of 2009 the registered decrease was 10.1 per cent. Also industrial production in the first six months of the year had fallen by almost 14.8 per cent compared with the same period in 2008, while investments had reached only 81.1 per cent of the level from the first six months of 2008. Exports for the first six months of 2009 were 47.4 per cent lower than in the same period of 2008. Even more disturbing was the increasing number of Russians who were now jobless – unemployment was 42.8 per cent higher than in the first six months of 2008.[1] Even those who had kept their jobs were hit by the crisis – according to World Bank figures, the average wage in Russia had fallen from USD 738 in December 2008 to USD 587 by May 2009.[2]

The year 2009 was the first year in the new century when Russian GDP fell, and proved to be one of the worst years in the entire modern economic history of the country, with worse economic performance noted in only three years – 1917, when GDP had shrunk by as much as 18.2 per cent, and two early years of Yeltsin's economic revolution – 1993 with a fall of 14.5 per cent, and 1995 with 12.7 per cent.

The crisis revealed various structural deficiencies of the Russian model of economic development and – even more worrisome – the country's overwhelming dependence on the global oil price and its vulnerability to global economic trends (Gaddy and Ickes, 2010).

The nine chapters of this book have been prepared by leading Western and Russian experts in the field of Russian energy policy studies. They cover central aspects of Russian energy policy after the 2008 crisis that have forced policy makers to reconsider their ideas and plans.

The Structure of this Book

Any study of the factors shaping Russian energy policy must take a range of issues into consideration. Energy policy is shaped by various actors and stakeholders, so the agency dimension of policy shaping is undoubtedly important. But the decisions and actions of those actors are in turn affected by many factors. How actors make decisions and how they behave is influenced not only by their position in the institutional and political landscape of Russia. Also their identity, their understanding of the world and of Russia's place in it, and their ideas on natural resources, on how the energy sector should be organized and interact with other

---

1   http://www.gks.ru/bgd/free/B09_00/IssWWW.exe/Stg/d06/1-0.htm

2   http://siteresources.worldbank.org/INTRUSSIANFEDERATION/ Resources/305499-1245838520910/rer19-eng_macro.pdf

sectors of the Russian economy and with partners, customers and competitors abroad – all these factors are important in shaping energy policy. Until recently, and due largely to Russia's being endowed with seemingly infinite energy resources, the issue of energy saving was not very high on the political agenda. The reasons are both economic and cultural: many Russians living today grew up in a system that did not encourage energy saving, and made access to inexpensive energy something that could be taken for granted.

Energy resources have played a part not only in the process of shaping the economic and social space in Russia, but also in shaping its foreign and security policy. This is a country that in the course of the past twenty-odd years has had to cope with grave political, social and economic problems while also striving to play a more important part on the international stage. We need to understand how Russian decision makers think about the role these energy resources may play in helping their country achieve its strategic goals both at home and on the international scene.

Energy policy does not unfold in a legal, political, economic, organizational, social, power or ideological vacuum. The exports of energy commodities generate huge revenues for the Russian state and for Russian companies, but they also render the country highly dependent on fluctuations in the global energy market. Cooperation in the field of energy is the backbone of Russia's relations with its most important partner, the EU – but, given the lack of developed infrastructure, especially in the gas sector, this also makes Russia vulnerable to external influences. Kremlin decision makers and policy makers must take into account various structural factors that they have only limited ability to influence but that may have a massive impact on whether they can achieve their ambitions, in energy and in politics. In this study, 'structural factors' are defined as those elements that constrain policy choices made by policy makers and may have impact on policy outcomes by functioning as a 'reality check'. Such structural factors can be economic, like the process of oil price formulation or balancing between energy demand and supply in both global and regional context; they may be geological, like natural resource endowment; geographical, like the distance between production sites and markets; infrastructural, like the existence – or not – of pipeline network to transport energy commodities from point A to point B. Further, they may be social factors, like the social capital of the country that realizes its energy policy and therefore needs a specific human competence; institutional, like the quality of institutions and administration that may prove decisive when the policies designed by policy makers are to be implemented; technological, like access to technology that makes it possible to produce energy from fields located offshore, or from fields with specific geological features, like those in which shale gas and oil are located. There is also the issue of relative power relations: this explains why a country like Saudi Arabia needs an army to protect its energy resources against external threats, making it one of the world's major arms purchasers; or why Norway, which lacks strategic capacities, decided to become a member of an alliance to strengthen the 'deterrence leg' of its policies.

The contributions to the book are indeed wide-ranging. There is a focus on the role of actors and their ideas in the shaping of Russian energy policy (Jakub M. Godzimirski), the resource base of the Russian energy sector (Valeriy Kryukov and Arild Moe), on how the sector has adapted to the post-crisis period (Tatiana Mitrova), on the impact of the crisis experience on the modernization discussion and the future role of energy in Russian strategy (Derek Averre), on the link between Russia's energy resources and its foreign policy (Irina Busygina and Mikhail Filippov), on initiatives for making the country's energy sector less dependent on supplies to Europe and transit areas (Pavel Baev), on the economic connection between Russia's energy exports and domestic reform of the energy market as exemplified by developments in the gas sector (Lars P. Lunden and Daniel Fjaertoft) and, finally, on future scenarios for the development of the Russian gas sector (Kari Liuhto, Enni Laaksonen and Hanna Mäkinen).

The specific topics have been carefully chosen in order to point up the various interconnections between areas not always treated as important in studies of Russian energy policy. Godzimirski looks at what could be labelled the 'power sociology' of the Russian energy sector; Averre explores how ideas on the need for modernization of Russia aired by some representatives of the country's political elite have become the name of the game that could – if played – change Russia and the Russian energy sector. Several other authors, like Kryukov and Moe as well as Mitrova, examine the structural factors that shape Russian energy sector – the energy resources without which Russia could not play such an important energy role, and the market conditions over which Russian policy makers have only limited power. Baev, as well as Busygina and Filippov, explore the spatial patterns of Russian energy actions by analysing how the Russian energy sector and its decision makers have been coping with the constraints that shape the spatial room for energy actions (Baev), or by examining how the fact that Russia's major economic and political partner, the EU, is dependent on energy supplies from Russia shapes the room for foreign policy making both in Russia and in Europe. The chapters by Lunden and Fjaertoft and by Liuhto, Laaksonen and Mäkinen focus on questions related to Russia's gas market, its future development and how this development will be influenced not only by market conditions but also by the political decisions made by policy makers.

The first issue discussed in detail is the question of agency in Russian energy sector. Who are the actors that influence its development? How may their ideas shape that important sector of the Russian economy? These questions are taken up in the first chapter, where *Godzimirski* maps the personal landscape of energy policy making in Russia under the presidencies of Vladimir Putin and Dmitrii Medvedev. The point of entry is the personal dimension of Russian energy sector – the composition of the main formal and informal bodies central in defining policy priorities and goals. Godzimirski examines their ideas on the development of the energy sector, and the efforts made to translate these into political action. The analysis focuses on where actors with stakes in Russian energy sector can be positioned on a 'reputational map' of informal political power, and on the composition of one formal body – the Governmental Commission on Fuel and the Energy Complex, the Mineral Resource Base and the Energy Efficiency of the Economy (the 'Sechin Commission') – where most actors representing state as well as private interests could meet to discuss the future of the country's energy sector. Godzimirski further details the process of reformulation of energy policy in the wake of the 2008 economic crisis which led, among other things, to a new reopening of the Russian energy sector to foreign oil and gas companies. The series of deals on cooperation between Russian and foreign oil and gas companies signed in 2011 and 2012, and especially the process that resulted in the deal on close cooperation between Rosneft and BP, show how ideas discussed by Russian policy makers and representatives of the leading global energy companies have been effectively translated into policy actions that are set to change the organizational landscape of Russian energy sector (see also Overland et al., 2012).

The ensuing chapters focus on various aspects of Russian energy policy and on the opportunities and constraints that influence the choices made by the country's energy decision makers. In the second chapter, *Valeriy Kryukov* and *Arild Moe* offer a detailed account of the state of oil reserves in Russia and of how to calculate these so as to enable a more realistic picture of what is still left under the ground. The issue of resources is crucial for understanding Russian energy policy: the key task of this policy is to determine how these resources can be used to cover domestic energy needs, how these resources should be marketed domestically and abroad, what to do with windfalls generated by those resources, and how to use these resources in pursuing Russia's goals in other areas. In that sense Kryukov's and Moe's chapter can be viewed as the foundation on which the whole book rests, because the remaining contributors deal with questions related to the issue of resources and their use in the economic or political context. Kryukov and Moe present the main lines in the Russian debate on resources versus reserves, noting how various ways of calculating reserves may create partly distorted pictures of the situation. They show how the Russian system of classifying reserves and resources may be misused to present a more 'optimistic' picture. They also discuss the main dilemmas faced by Russian energy decision makers – not least the uneven geographical distribution of reserves, the quality of the existing reserves and size of new additions. They examine how Russian oil producers can increase production from existing oil fields by using new technology and how stepped-up drilling has not resulted in new oil discoveries. And finally, the authors discuss the response of various actors and the future of Russian oil sector, including the role of projects in the Russian Arctic.

In the third chapter, *Tatiana Mitrova* discusses how Russian energy sector has been adapting to the new post-crisis market conditions in a situation with lower demand and the end of the era of reaping low-hanging energy fruits in Russia. She takes up a range of critical questions, including the impact of the 2008 crisis on the energy sector and how the Russian expert community was compelled by the impact of the crisis to reconsider the goals and means of the country's energy strategy until 2030. Mitrova presents the main goals of this new reformulated strategy, which include a greater more focus on national interests in the emerging international regulation system, more interest in diversification of exports and export markets, diversification of export routes and securing stable market conditions and reasonable prices for Russian energy commodities. She also examines issues like the position of Russian energy companies abroad and the importance of international energy cooperation in developing difficult energy projects, for instance in the Arctic. She then goes on to focus on specific issues linked to the development of Russian oil, gas and coal industry. She rounds off the chapter with several conclusions, notably that realization of Russia's energy policy in the post-crisis period is a highly challenging task, and that Russia has been only partly successful in adapting to those new changing circumstances. Mitrova lists a number of challenges Russian decision makers will have to deal with if Russia is to retain its central role as a major exporter of energy: especially important is the need for new technology, better project management and putting in place a new and more consistent regulatory framework for energy-sector operations.

Mitrova's chapter is very much about the planned use of energy resources, which links it directly with the chapter by Kryukov and Moe, but she also looks at the external dimensions of Russian energy policy, painting a broader picture that is filled with more detail and content in the chapters by Baev, and by Busygina and Filippov. What Mitrova presents is also a good illustration of how energy policy in Russia has been reactive to global developments, which says a lot about the process of policy making – also a topic dealt with by Godzimirski in chapter one.

In chapter four, *Derek Averre* looks at the relation between the Russian debate on modernization and developments in the Russian energy sector. He identifies several parameters, including the state's influence over petroleum sector, the use of regulatory instruments to control energy assets, the distribution of privileges and, importantly, constraints placed on foreign investors, as a way of measuring the progress in modernization. Averre focuses on three questions – the motives underpinning Medvedev's modernization programme; how oil and gas resources are factored into achieving progress in modernization; and finally the issue of governance in the hydrocarbons sector. In particular, he examines whether current policies will allow the energy sector to become the locomotive of Russia's development. The relationship between the modernization programme and the country's energy resources has become pivotal, especially after the economic crisis revealed deficiencies of the prevailing economic policy, and not least Russia's interdependence with global structures. As the export of energy resources represented at its peak almost 70 per cent of Russia's exports, energy stands out as the most important element linking Russia in economic but also in political terms with the outside world. As the goal of modernization programme launched by Medvedev in the wake of the crisis in 2009 was to improve Russia's international competitiveness, and since energy commodities are the major element of its economic relations with the world, successful implementation of the modernization plans will necessarily have direct consequences for the fate of the country's energy sector. Averre examines how modernization has become an issue on Russia's foreign policy agenda, as well as the role of energy sector in the work on modernization. Genuine modernization of the Russian energy sector is an important part of the modernization plan, as only a modernized energy sector will be able to secure stable and sufficient deliveries of energy commodities to the domestic market and for export. If the Russian energy sector is to be able to exploit dwindling or increasingly difficult-to-access sources of hydrocarbons, upgrade processing and pipeline facilities, and introduce clean energy and energy-saving technologies, then close cooperation with foreign investors willing to share technological, organizational and economic risks with their Russian partners will be crucial. Such cooperation can in turn contribute to modernization not only of Russian energy sector but also the country as a whole.

Averre focuses on how the ideas of various actors influence Russian policy making and the choice of modernization path. This complements Godzimirski's chapter, showing how those actors' ideas influence – or do not – the development of the Russian energy sector. By looking into the importance of modernization agenda in foreign policy, Averre's chapter builds a bridge to the next chapters, which focus on various aspects of the international energy market games in which Russia is involved.

In chapter five, *Irina Busygina and Mikhail Filippov* offer a theory-based explanation of why cooperation between Russia and the EU has remained so severely limited in many areas of mutual interest, especially as regards policies in the post-Soviet region. They study the indirect and less obvious consequences of the close energy cooperation between Russia and the EU, showing that this cooperation has not had solely positive impacts on relations between those two important centres of political and economic power. They see the deterioration of relations between Russia and the EU as the result of a strategic choice, and as a function of high energy prices and the growing importance of energy trade between the two actors. They find that the high share of revenues from export of energy commodities in the Russian economy has created the wrong incentives not only in the Russian economy, but also in the country's foreign policy. Further, they explain, this deterioration of relations was needed in order to give some legitimacy and maintain the hybrid regime at home. The creation of 'virtual conflict' in relations between Russia and the EU, and Russia and some post-Soviet countries, also presented to the Russian political audience as Russia's conflict with the EU for influence in the post-Soviet space, has played a crucial role in the consolidation of the current regime, without forcing Russia into economic and informational isolation. Because the West needs Russian energy resources to meet its energy needs, this virtual conflict does not damage Russia's economic relations or its energy exports to the West, which continue to secure constant inflows of revenues. Busygina and Filippov describe this situation as a counterintuitive trade-off, seeing improved economic relations as leading to more political conflict, not less. This paradoxical development is due largely to the use of 'diversionary tension' which they see as a specific feature of Russian foreign and domestic policy making, employed especially in relations with Ukraine and during the two 'gas wars'. To the domestic audience, the leadership presents its foreign policy as Russia's competition with the West in the post-Soviet space, while Russia continues to reap huge economic benefits from its energy trade with the EU. This chapter clearly shows how the fact of being a main producer and exporter of energy shapes not only Russia itself but also its relations with the outside world.

Russia's dependence on European energy market has been the major driving force behind attempts at diversifying not only transit routes but also energy markets. Those questions are dealt with by *Pavel Baev* in chapter six, on Russia's search for security of demand and transit. Baev explores how the Russian state and Russian companies have been striving to make Russia less dependent on sales of energy to Europe and on transit countries, in particular Ukraine and Belarus. He begins by introducing the Russian version of the debate on diversification, and follows up by presenting Russia's recent diversification efforts – in West, South and East. He details the various steps taken recently and assesses the successes and failures of Russian policies. Paying special attention to energy relations with countries that figure high on the Russian energy agenda – Germany, Belarus, Ukraine, Turkey and China – Baev shows that there is still a gap between Russian declarations and actual achievements. Further, he explores the state of energy relations between Russia and the EU, with Russian reluctance towards the Energy Charter Treaty, and notes the legal battles between Russia and the EU. His central focus is on the construction of energy infrastructure that can make Russia less dependent on transit countries, first and foremost the construction of the Nord Stream and the South Stream pipelines. As these pipelines will have severe consequences for Belarus and Ukraine, Baev examines energy relations between those two countries and Russia, not least their attempts at persuading Russia to adopt a policy less detrimental to their own energy interests. He then turns to Russia's emerging energy relations with China, presenting the considerable complexity of this issue.

Baev's main conclusion is that the Russian state has evident problems with strategic coordination of its energy policies, as shown by the fact that the development of the new transport infrastructure in the form of gas and oil pipelines has not been coordinated with the development of greenfields. This may in turn mean that Russia is on the verge of building a new and costly infrastructure that will be only partly used. These conclusions would indicate that Russia is suffering from too little strategic coordination rather than too much – an interesting point, in light of the ongoing debate on Putin the Almighty. Perhaps we will have to revise our assessment of the efficiency of the current Russian regime and its ability to pursue long-term strategic goals cost-efficiently. Or it might mean that the regime is willing to pay a relatively high price for achieving its mostly political goals.

Whether the regime is vague and inconsistent is explored in chapter seven. Economists *Lars P. Lunden* and *Daniel Fjærtoft* examine the Russian debate on introducing market prices for gas in Russia, asking what consequences such a decision could have for the availability of additional volumes of Russian gas for export. The decision on whether Russian gas consumers should pay the same price for the gas they use as their European counterparts is primarily a political one, but the consequences will be both political and economic. Lunden and Fjærtoft explore whether higher domestic gas prices might lead to increased Russian export of gas. Their analysis indicates that unless domestic price hikes are accompanied by reforms in other areas, there will be not much more gas available for export. Moreover, price reform might have a side-effect, offering further incentives for Russia to use gas exports as a tool of foreign policy. Lunden and Fjærtoft employ a simple macroeconomic framework in seeking answers to four key questions – three from the realm of economy, and one on the nature of the relationship between Gazprom and the Russian state. The three economic questions concern the impact of price reform on availability of gas for export, the impact of gas price reform on investments in production, and the incentives the current gas price reform may give to Gazprom to re-allocate gas supplies. The political question concerns the impact of gas price reform on the possible use of gas as a political weapon in the pursuit of foreign policy goals. The authors discuss price reform in a broader political and economic context, showing how the decisions on its implementation – or rather lack of implementation – have been influenced by the ratings of political leaders, by the volatility of gas price on the European market and by the effects of the crisis on Russia. They also note a paradox – the planned introduction of market-based price reform in a situation when the gas price in Russia is in fact decided not by market forces but by officials who tell Gazprom what price the company can charge and how much gas is to be supplied to domestic consumers. To get a better understanding of the effects the price reform could have on the availability of gas for export, Lunden and Fjærtoft undertake a multi-variable analysis of the impact of various economic factors like price elasticity, income, and energy savings on this process. All in all, they paint a complex picture of factors that influence developments in Russia's gas sector: the sector remains dominated by the state-controlled Gazprom, but Gazprom finds itself challenged by a range of 'independent' gas producers whose interests often collide with those of that state-owned gas behemoth. Gazprom has to adapt to quickly changing circumstances: and that is proving to be a challenging task, not least in the domestic environment where the company's economic interests must often yield to the political interests of the ruling elite. Also on the European market Gazprom faces dilemmas, like finding a balance between supplying this market with higher volumes of gas without putting pressure on the price of that commodity.

The question of international gas market and Russia's future position is taken up in chapter eight, co-authored by *Enni Laaksonen, Hanna Mäkinen* and *Kari Liuhto*. They present a detailed analysis of the current and future trends influencing the development of regional and global gas markets, including the issue of shale gas development. They then link together various aspects of that development to present a more nuanced picture of the setting in which Russian gas policy is to be realized. As their focus is mainly on gas, this chapter complements that of Kryukov and Moe, which focuses more on oil resources. Chapter eight also represents a useful supplement to Baev's chapter, with its discussion of Russian external energy strategy in three directions, and to Lunden's and Fjærtoft's chapter seven, with its analysis of a more theory-based econometric framework for Russian gas policy. Chapter eight can also be viewed as a comprehensive account of how the Russian gas strategy as described in the Energy Strategy until 2030 – presented by Mitrova in chapter three – has been actually realized and what future challenges Russian policy makers may face. In addition, the three authors outline the major resource-related challenges facing the Russian gas sector, like falling production from the existing megafields and the need to develop new resources, first and foremost the Yamal Peninsula and the Shtokman gas field in the Barents Sea. In the second part of the chapter they examine the prospects for Russian gas exports to Europe, Asia and the USA, and the impact that the development of unconventional gas resources in Europe and in Asia (China) might have on Russia's future position on the regional and global gas markets. The authors hold that Russia faces today a difficult gas future – according to one of the scenarios discussed, the market may face a shortage of gas and therefore high prices for that commodity; but in another, the supply of gas may exceed demand on the European market, leading to lower gas prices. In order to avoid these extremes, Russia and Russian companies will have to embark on a balanced policy of field development combined with skilful management of gas resources.

In the ninth and final chapter, the editor of this volume, *Jakub M. Godzimirski*, sums up the findings presented in the preceding chapters. Viewing the possible and impossible futures of the Russian energy sector in a broader strategic perspective, he presents the most recent steps taken by Russian policy makers after Putin's return to the presidency. Here he offers some insights into how the realization of Russia's grand energy strategy may be obstructed because the framework conditions are shaped by a combination of structural factors where Russian policy makers have only limited influence. A central issue discussed throughout this book concerns how the Russian state – and its ruling elite – must deal with factors and forces that are partly or completely beyond their control.

In that sense, this volume follows the International Political Economy (IPE) tradition where a key theme is the relationship between state policies, including energy policies, and international markets (Gilpin, 1987, Gilpin, 2001). In his earlier work, Gilpin presented interesting thoughts on relation between state and market (1987: 10–11). He noted that the state and the market cannot be totally separated even if they are based on completely different principles – the state on the concepts of territoriality, loyalty, exclusivity, and monopoly of the legitimate use of force; and the market on the concepts of functional integration, contractual relationships, and expanding interdependence of buyers and sellers. Further, according to Gilpin, the state sees territorial boundaries as a necessary basis of national autonomy and political unity – whereas for the market, eliminating all political and other obstacles to the operation of the price mechanism is imperative. The tension between these two fundamentally different ways of ordering human relationships has shaped the course of modern history: Gilpin sees them as constituting the crucial problem in the study of political economy.

That same tension also emerges as an important factor shaping Russian energy policy, which is the main topic of this volume. What makes the study of the relationship between the Russian state and the international energy market even more interesting – and challenging – are the specific features characteristic of both the Russian state and the international energy market. We sincerely hope that this volume will help readers to understand the complexity of Russia, the complexity of issues in the international energy market, and, perhaps the most challenging – the complexity of the relationship between Putin's Russia and the forces of the global energy market.

References

BP 2012. Statistical review of world energy 2012. Available at http://www.bp.com/sectionbodycopy.do?categoryId=7500&contentId=7068481 [accessed 10 November 2012].

Gaddy, C. and B. Ickes 2010. Russia after the global financial crisis. *Eurasian Geography and Economics*, 51(3), 281–311.

Gaddy, C. 2011. Will the Russian economy rid itself of its dependence on oil? Available at http://valdaiclub.com/economy/26960.html [accessed 12 December 2012].

Gilpin, R. 1987. *The Political Economy of International Relations*. Princeton, NJ: Princeton University Press.

Gilpin, R. 2001. *Global Political Economy: Understanding the International Economic Order*. Princeton, NJ: Princeton University Press.

Hill, F. 2004. *Energy Empire: Oil, Gas and Russia's Revival*. London: Foreign Policy Centre.

IISS [International Institute of Strategic Studies] 2006. Russia: an emerging energy superpower? in *IISS Strategic Survey 2006*. London: IISS.

Medvedev, D. 2009. Go Russia! Moscow: Kremlin. Available at: http://archive.kremlin.ru/eng/speeches/2009/09/10/1534_type104017_221527.shtml [accessed 10 November 2012].

Ministry of Energy, Russian Federation 2011. Energodialog Rossiya - ES. Available at http://minenergo.gov.ru/upload/iblock/987/9878cea29d281a7168f1d0d2668ebbb0.pdf [accessed 15 November 2012].

Ministry of Energy, Russian Federation 2012. Statisticheskaya informtasiya Available at http://minenergo.gov.ru/activity/statistic/10478.html [accessed 15 November 2012].

Orttung, R. W. and I. Overland 2011. A limited toolbox: explaining the constraints on Russia's foreign energy policy. *Journal of Eurasian Studies*, 2(1),74–85.

Overland, I., J. Godzimirski, L. P. Lunden and D. Fjaertoft 2012. Rosneft's offshore partnerships: the re-opening of the Russian petroleum frontier? *Polar Record* Available at http://journals.cambridge.org/action/displayAbstract?fromPage=online&aid=8526083 [accessed 23 August 2012].

Rutland, P. 2008. Russia as an energy superpower. *New Political Economy*, 13(2), 203–210.

WTO [World Trade Organization] 2011. World Trade Report 2010. Trade in Natural Resources. Available at http://www.wto.org/english/res_e/publications_e/wtr10_e.htm [accessed 22 November 2012].

Chapter 1
Actors, Ideas and Actions

Jakub M. Godzimirski

Introduction

This chapter presents the key actors, ideas and actions shaping the Russian energy sector in recent years. The analysis focuses on three closely intertwined aspects: the actors who shape Russian energy policy, their ideas and world outlook, and the most important actions they have undertaken in the post-crisis period (2008–2012). Which actors have been central to the processes of formulating and implementing of energy policy in Russia after the crisis? Has the crisis led them to reformulate their policies and goals? Since Tatiana Mitrova's chapter in this book examines how the crisis has affected the formulation of goals in Russian energy policy, this chapter focuses on the actors, their ideas and their political actions.

We begin with a look at the key figures, followed by a brief analysis of their ideas on the role of the energy sector in Russia. The third part looks at their main actions that have shaped this important sector of the Russian economy.

Actors

There are at least three possible approaches to identifying the key actors in the Russian energy sector. The first is to consult the lists of owners and managers of major energy companies to see who is in charge of development.[1] That would rapidly reveal that the state is an important owner, controlling approximately 80 per cent of gas production, between 30 and 40 per cent of oil production, and with an effective monopoly on transport and export of those two major energy commodities.

This clearly dominant role of the Russian state in the energy sector may make the second approach to identifying key actors more viable. That approach would use the various Russian rankings of political, reputational and economic power published regularly in the media – such as the rankings of the country's top 100 politicians published by *Nezavisimaya Gazeta* – to identify key actors in the energy sector.

---

1 For such an overview see the list of rankings of the most influential owners and managers in the Russian energy sector published monthly by *Neft Rossii* and available at http://www.oilru.com/articles/.

The third way would be to analyse the composition of the state bodies responsible for the development of the energy sector. Which energy actors have a seat in those official bodies and thereby a direct say in shaping energy policy in Russia?

Here, however, it should be borne in mind that the Russian political system is characterized by several specific features. The energy sector has been defined as 'strategically important'– and strategic decisions require not only economic justification on the company level, but also political approval at the highest level. For that reason, we will combine the three approaches. Only by comparing and merging various 'name lists' can we see whose influence has a visible impact on the development of Russia's energy sector.

This analysis is based on a top–down approach. The truly strategic decisions are probably taken informally at the highest political level and then 'pushed down' the power vertical. That does not mean that those decisions will always be implemented smoothly, as various groups of interests or individual actors may block their realization. A clear illustration here could be the fate of the strategic deal on cooperation and assets swap made between Rosneft and BP in January 2011 that was directly supported by Vladimir Putin and Igor Sechin: it was effectively 'killed' – at least temporarily – by a group of powerful Russian oligarchs led by Mikhail Fridman.

This painful reputational defeat notwithstanding, the 'Putin team' still seems to be very much in charge. Even though their party of power, United Russia, failed to cross the 50 per cent threshold in elections to the new State Duma on 4 December 2011, and a wave of protests swept the major cities, the opposition was apparently too weak to pose any real threat to the ruling group.

This impression was also confirmed by the annual ranking of political power published by *Nezavisimaya Gazeta* on 16 January 2012 (Orlov, 2012), and then by the results of presidential elections won by Vladimir Putin. Summing up the results of the year 2011, Orlov wrote that Putin had preserved and even strengthened his position; that leaders working closely with Putin had improved their positions; and that the members of the current elite had consolidated their grip on power. The federal administrative elite dominated the ranking, followed by members of the party elite, the regional elite and the business elite. Almost 60 per cent belonged to the first group, 20 to the second, with the regional and business elite represented by 10 per cent each.

The same five names occupied the top five places in 2011 and in 2010, but with some important changes. In 2010 the Putin–Medvedev tandem had been followed by Aleksander Kudrin (Minister of Finance), Vladislav Surkov and Igor Sechin. In 2011 the same tandem was followed by Surkov, Sechin and Kudrin, in that order. In 2012 a clear regrouping on the top of the Russian power pyramid took place – Putin and Medvedev retained their top positions, but they were followed by I. Shuvalov, V. Volodin and S. Ivanov. Igor Sechin was demoted to the sixth place, Surkov – to the eight place, while Kudrin ended up at 77th place.[2]

---

2   http://www.ng.ru/ideas/2013-01-14/9_top100.html.

This demotion notwithstanding, Aleksander Kudrin has indeed played a crucial role in shaping the framework conditions for the energy sector since 1996, when he joined the Presidential Administration as its deputy head, working closely with Anatolii Chubais. It was Kudrin who proposed that Putin be appointed as deputy head of Yeltsin's presidential administration, when he himself left the administration to work as Russia's representative in the International Monetary Fund in May 1997. In the same year Kudrin was appointed state secretary in the Russian Ministry of Finance. After a brief break between January and June 1999 he was reappointed as first vice-minister of finance and in May 2000 was appointed Minister of Finance of the Russian Federation. He served continuously in that capacity until September 2011, when he was forced to resign. During his time at the top of Russian politics he played a crucial part when decisions were made on increasing the fiscal burden for Russian energy producers by reintroducing (in 1999) export duties and linking the level of taxation to global energy prices, thereby securing additional revenues for the state budget. Further, the introduction of the severance tax, known in Russia as the Mineral Extraction Tax (NDPI) in 2002, replacing a more complex earlier tax system, was to provide the state with additional mining rent (Goldsworthy and Zakharova, 2010). This focus on the fiscal aspects of energy policy also resulted in the budget proposal made in 2006, where Kudrin proposed the establishment of the Russian Stabilization Fund, which helped the country to cope with the financial crisis and made the state budget better prepared to cope with the volatility of oil prices by providing a financial cushion.[3]

However, Kudrin's power position was drastically weakened by his row with Medvedev, which finally resulted in his dismissal from the Russian government in late September 2011. While in August 2011 Kudrin had occupied the third position in the Top 100 ranking, already by October, only a few weeks after his argument with Medvedev, he had been demoted to 47–49 position,[4] and by November 2011 his name was in 77th place on the list.[5] He managed a slight recovery in December 2011 (returning to 44th position) – due mainly not to his own achievements but to the fact that Putin had mentioned publically that Kudrin was still to be treated as a member of his team.[6] In the meantime, Anton Siluanov, Kudrin's former deputy who was appointed Minister of Finance after Kudrin's forced departure, not only made the list in November 2011 when he was ranked 93, but soared to 28–29th position by December 2011 and was expected to enter the top ten in the first months of 2012.

In the same period, Igor Sechin, who was formally responsible in the Putin government for the political management of the country's energy sector, climbed from fifth position in August 2011 to third position in December 2011, overtaking not only Kudrin, who had fallen from grace, but also Vladislav Surkov, the

3 Lentapedia: Aleksey Kudrin: http://www.lenta.ru/lib/14161127/full.htm.
4 http://www.ng.ru/ideas/2011-11-01/6_top100.html.
5 http://www.ng.ru/ideas/2011-11-30/9_top100.html.
6 http://www.ng.ru/ideas/2011-12-29/5_top100.html.

influential deputy head of the Presidential Administration and a man seen as the main ideologue of the Kremlin. Sechin's formal appointment as Deputy Prime Minister responsible, *inter alia*, for the development of nation's energy sector,[7] his friendship with Vladimir Putin dating back to the early 1990s and his ability to play a major part in the current political game made him one of the three most influential players in Russian politics (Sakwa, 2011; Reznik and Mokrousova, 2011). According to some observers, he was at some stages even more influential then President Dmitrii Medvedev[8] (see also Pribylovskiy, 2010: 5–6). In any case, there can be no doubt that Sechin's formal and informal power has been much greater than that of any other energy player in Russia.

From the 2011 list of Russia's most powerful figures it was clear that many of them played a role or had a stake in the energy sector, whether as state managers and state representatives (Miller, ranked as no. 17; Zubkov, 29; Tokarev, 59; Kiriyenko, 72), as owners (Alekperov, 25; Usmanov, 34; Fridman, 57; Yevtushenkov, 69; Aven, 95), as traders (Timchenko, 35), or as policymakers (Kudrin, 5; Shuvalov, 10; Trutnev, 66; Shmatko, 85).

Many of them have also sat on the Governmental Commission on Fuel and the Energy Complex, the Mineral Resource Base and the Energy Efficiency of the Economy under Deputy Prime Minister Igor Sechin.[9] The main tasks of the Commission were to coordinate cooperation among federal executive bodies, executive bodies of subjects of the Russian Federation and other organizations, to work for sustainable development and operation of the country's fuel and energy complex, and to ensure the development and implementation of state energy policy. The list of members and their institutional and business affiliations can therefore serve as a useful guide showing who the key players in the energy sector were during Medvedev's presidency and while Sechin was formally responsible for shaping Russia's energy policy. The Commission had grown from 32 members in 2010 to 36 by 2011. Fourteen of these members represented various state institutions, two represented the legislature, one represented regional elite, and the remaining 19 came from the business community.

Although the business community may appear well represented, we should recall that the Russian state is a key owner in many companies represented in the Commission: in Rosneft (via Rosneftegaz) in 2011 the state owned 75.16 per cent; in Gazprom Neft (via Gazprom) 95.7 per cent; in Gazprom 50.002 per cent; in Transneft 78.1 per cent (the state's share of votes here is 100 per cent); in Holding MRSK 53 per cent; in the Federal National Electric Grid 79.11 per cent; and in the Russian Railroads 100 per cent. In addition, Rosatom is organized as a state corporation. The composition of this important Commission thus confirms the dominant position of the Russian state in the energy sector.

---

7   http://www.government.ru/gov/activity/#person8.

8   http://www.forbes.com/lists/2009/20/power-09_Igor-Sechin_XIE4.html.

9   www.government.ru/gov/agencies/143.

Table 1.1 combines data from the 12 annual Top 100 rankings published by *Nezavisimaya Gazeta (NG)* between 2001 and 2012 with an overview of affiliations of members of the Sechin Commission. The names are presented with those with greatest political influence in 2011 at the top and those with less political clout at the bottom. This should offer a good indication of the most influential energy actors as of the end of 2011, and in which institutional environment they operated, representing the interests of key energy companies and of various state institutions with direct or indirect stakes in the Russian energy sector.

Several members of the Commission deserve further mention. Among them is Boris Kovalchuk. Although he does not feature in any of the *NG* Top 100 rankings in the period studied here (2001–2012), he has a strong family connection to the Russian power elite – his father, Yuriy Kovalchuk, co-owner and manager of the Rossiya Bank, is a close associate of Vladimir Putin, with whom he became acquainted in St Petersburg in 1991. His close association with Putin made him an important political figure in Putin's Russia, and his name featured in all *NG* Top 100 rankings since 2007, with best position achieved in 2008 (ranked as 27–28) and the lowest in 2011 (88).

Anton Ustinov, who is thought to be a relative of Vladimir Ustinov, a close ally of Sechin (Sakwa, 2010, p. 188), served as the head the legal department of the Federal Tax Service. In February 2008, Kudrin fired him from that post, but in May 2008 he was appointed aide to Deputy Prime Minister Igor Sechin. According to the Russian media, Ustinov has played a key role in many of the most notorious tax cases in the country, including the Yukos case (Vedomosti, 2008; Ukhov, 2008).

Eduard Khudainatov is another interesting case. He had a seat on the Commission as the CEO of the state-owned oil and gas company Rosneft. In 2010 he replaced Sergei Bogdanchikov, who was believed to be Sechin's man and who managed to use his position to gain political clout. Bogdanchikov had been ranked among the most influential figures in Russia in the years between 2004 and 2010, climbing to 28th position in 2006, then falling to 87th position in 2010 and disappearing from the list in 2011 after he had lost his post in Rosneft.

A member of the Commission with apparently increasing political clout is Leonid Mikhelson, CEO and main owner of Novatek, the biggest of the 'independent gas producers' in the country and close associate of another important Russian/Finnish energy player, Gennadii Timchenko, who is also one of Novatek's main shareholders. In early 2012, it was even rumoured that Mikhelson might replace Alexei Miller as the CEO of Gazprom.[10] In that case, Mikhelson would probably join the group of the most influential energy policy makers in Russia. Regardless, he is still among the major energy players in the country,[11] and the one who managed to break the Gazprom export monopoly effectively, by signing the deal with Total on production and export of LNG from Yamal.[12]

---

10   http://lenta.ru/news/2012/01/20/fired/.
11   http://lenta.ru/news/2011/02/15/ceos/.
12   http://lenta.ru/news/2011/03/03/up/.

Table 1.1 Members of the Sechin TEK Commission, ranked by their position in the *Nezavisimaya Gazeta* top 100 ranking 2011

Name	Function as listed in Sechin's Commission	State official, political position	State controlled company, organisation	NG Top 100 2000–2011 occurrences	NG Top 100 2000–2011 Range Top – Bottom	NG Top 100 2011[†]
Sechin, I.	Deputy Prime Minister (Chairman of the Commission)	x		12	3 (2011) – 72 (2001)	3
Miller, A.	Chairman of Gazprom		x	11	9 (2007/2008) – 70–71 (2001)	17
Alekperov, V.	President of LUKOIL		x	12	14 (2006) – 31 (2008)	25
Yakunin, V.	President of Russian Railways		x	6	13 (2007) – 43 (2011)	43
Levitin, I.	Minister of Transport	x		8	53 (2011) – 99–100 (2004)	53
Tokarev, N.	President of Transneft		x	5	59 (2011) – 85 (2007)	59
Trutnev, Y.	Minister of Natural Resources and Environmental Protection	x		8	51 (2008) – 75–77 (2004)	66
Yevtushenkov, V.	Chairman of the Board of Directors of Sistema			8	34 (2005) – 78 (2010)	69
Belyaninov, A.	Head of the FCS of Russia	x		4	60 (2009) – 84 (2007)	76
Shmatko, S.	Minister of Energy (Vice Chairman)	x		4	74–76 (2009) – 90 (2008)	85
Bogdanov, V.	General Director of OJSC 'Surgutneftegas'			2	58 (2006) – 58–59 (2005)	
Budargin, O.	Chairman of Federal Grid Company UES		x	0		
Dod, E.	Chairman of the Federal Hydro Company			0		
Dyukov, A.	General Director of Gazprom Neft		x	0		
Golomolzin, A.	Deputy Head of FAS Russia	x		0		
Khan, G.	Executive Director of TNK-BP Management			0		
Khudainatov, E.	President Rosneft		x	0		
Komarova, N.	Governor of Khanty-Mansiysk Autonomous Okrug	x		0		
Kovalchuk, B.	Chairman of INTER RAO UES		x	0		

Name	Position			
Kudryashov, S.	Deputy Minister of Energy of the Russian Federation	x		0
Kutin, N.	Head of Rostekhnadzor	x		0
Ledovskikh, A.	Head of Rosnedr	x		0
Lipatov, Y.	Chairman of State Duma Committee on Energy	x		0
Lokshin, A.	Director of the Directorate of Nuclear Energy Complex Rosatom		x	0
Makarov, I.	Chairman of the Board of Directors of Itera			0
Maslov, S.	President of Saint Petersburg International Mercantile Exchange			0
Mikhelson, L.	Chairman, Novatek			0
Novak, A.	Deputy Minister of Finance	x		0
Novikov, S.	Head of the Federal Tariff Service	x		0
Ponomarev, D.	Chairman of the Council for Organizing Efficient System of Trading on the Wholesale and Retail Electricity and Capacity Market			0
Shvets, N.	General Director, MRSK Holding		x	0
Takhautdinov, S.	General Director, Tatneft			0
Tugolukov, E.	Chairman of State Duma Committee on Natural Resources	x		0
Ustinov, A.	Deputy Head of the Secretariat of Deputy Prime Minister Sechin II (Executive Secretary)	x		0
Voskresensky, S.	Deputy Minister of Economic Development	x		0
Yakovlev, Y.	Head of FSB Branch	x		0

Note: †http://www.ng.ru/ideas/2012-01-16/9_top100_2011.html.

Although the Commission was made up of the key players in Russian energy sector, it was widely seen as playing only a coordinating role, with the key strategic decisions being taken either by the Putin/Medvedev/Sechin trio or by the Putin/Sechin duet (Pribylovskiy, 2010; Kryshtanovskaya, 2011). Bearing in mind the highly hierarchical and personalized nature of politics in today's Russia, and in order to explore how these three key actors understand the role of the nation's energy sector, we now turn to their public statements and actions, which, it may be assumed, have played a major part in shaping Russian energy policy.

Ideas

The following analysis of statements by Putin, Medvedev and Sechin on energy in the Russian political context focuses on their public statements of diagnostic character that were conveyed to a broader Russian and international public between 2000 and 2011. Because Putin returned as president in March 2012, special attention will be paid to his views on the role of energy in Russia. Our focus is on public statements: less-known and less-read statements, such as Putin's and Sechin's doctoral dissertations and other academic writings discussed for instance by Balzer (2005; 2006), are not included here.

In order to identify some broader strategic long lines in both Putin's and Medvedev's approach to energy, an energy-centred analysis of their major public statements has been conducted. We examine whether and how energy-related issues have been discussed in Putin's 1999 'manifesto', Putin's seven addresses to the Federal Assembly (AFA) (2000–2007) and four of Medvedev's addresses to the Federal Assembly (2008–2011). In addition, an attempt has also been made to identify what could be termed Sechin's 'programmatic statements' on energy strategy. The choice of texts can be justified by their importance in the process of presenting the two leaders' strategic visions for the country. With Sechin, it was of interest to identify his main ideas on energy during his time as the main formal – and informal – political manager of Russian energy affairs, when he headed the TEK Commission (2008–2012).

When Putin was appointed Russian prime minister and before he was anointed heir apparent to Yeltsin, he (or his team) published what could be described as his first serious attempt at presenting his vision of Russia and of the challenges at that time (Putin, 1999). In this lengthy document there are only marginal references to energy-related issues, as when he promises that his government will pursue an economic policy of priority development of the leading industries in research and technology, and that one of the measures to be taken will be to buttress 'the export possibilities of the fuel and energy and raw-materials complexes'. In addition, he claims that Russia has had to pay a high price for the Soviet economy's excessive focus on the development of the raw materials and defence industries, which negatively affected the development of consumer production and services; and he promises to focus on the development of non-raw material industries as a way of addressing the needs of the Russian population.

After being elected president, Putin presented his views on various issues on many occasions. In particular, he used the annual addresses to the Federal Assembly (AFA) to present his views on Russia's needs and challenges, and most crucial problems facing the country and its political elite.

Table 1.2   Putin's and Medvedev's main analysed statements

Date	Full name of the text	Reference
1999 12 31	Putin: Russia at the Turn of Millennium Manifesto	1999 Manifesto
2000 07 08	Putin: Address to the Federal Assembly	2000 AFA
2001 04 03	Putin: Address to the Federal Assembly	2001 AFA
2003 05 16	Putin: Address to the Federal Assembly	2003 AFA
2004 05 26	Putin: Address to the Federal Assembly	2004 AFA
2005 04 25	Putin: Address to the Federal Assembly	2005 AFA
2006 05 10	Putin: Address to the Federal Assembly	2006 AFA
2007 04 26	Putin: Address to the Federal Assembly	2007 AFA
2008 11 05	Medvedev: Address to the Federal Assembly	2008 AFA
2009 11 12	Medvedev: Address to the Federal Assembly	2009 AFA
2010 11 30	Medvedev: Address to the Federal Assembly	2010 AFA
2011 12 22	Medvedev: Address to the Federal Assembly	2011 AFA

Source: www.kremlin.ru.

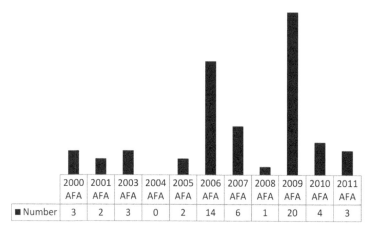

	2000 AFA	2001 AFA	2003 AFA	2004 AFA	2005 AFA	2006 AFA	2007 AFA	2008 AFA	2009 AFA	2010 AFA	2011 AFA
■ Number	3	2	3	0	2	14	6	1	20	4	3

Figure 1.1   Putin's and Medvedev's focus on energy in addresses to Federal Assembly – number of mentions of the term 'energy'
Source: www.kremlin.ru.

Putin's Ideas on Energy: AFA 2000–2007

Analysis reveals that Vladimir Putin mentions 'energy' a total of 30 times in his seven AFA speeches between 2000 and 2007. In only one of those speeches (2004) is energy in general not discussed at all, although he refers to 'creative energy' of Russians and pays some attention to developments in gas and oil sector; in his 2006 speech Putin mentions energy 14 times, but in the remaining five, energy-related issues are addressed only marginally. What types of energy have preoccupied Putin, and which energy-related areas are not taken up? We find 17 references to oil (one in 2001, two in 2003, four in 2004, one in 2006, and a full ten in 2007) and 16 references to gas (two in 2002, one in 2003, five in 2004, one in 2006 and seven in 2007 – seven), but coal and hydro energy are mentioned only three times each (both in 2007). In addition there are 26 references to nuclear energy, but mostly in the context of military and security policy, in connection with the importance of nuclear weapons in national security policy (2004 and 2006, with 14 references to the military uses of nuclear energy). Only in his 2007 address does Putin mention nuclear energy (seven times) as an energy source for the Russian economy. He also pays some attention to the development of the electricity and power generation sector (one mention in 2003, one in 2004 and ten in 2007). What is striking is Putin's lack of interest in renewables (although timber and peat are mentioned as potential sources of energy in 2009), and the complete absence of linkage between energy and climate change. He mentions 'climate' 11 times, but the focus is almost exclusively on the investment climate in Russia, not on climate change.

In his 2000 AFA speech Putin underlines the lack of competitiveness of Russian companies and notes that they have been able to survive thanks, *inter alia*, to access to low-priced energy. In this speech he also criticizes the continued orientation of the economy towards the raw materials sectors and the dependence of the state budget on revenues from the sales of energy resources on unstable global markets. He adds that the Russian state has aimed at creating equal conditions for competition and that some companies have been in a privileged position by being able to enjoy lower energy tariffs. Then he goes on to say that undeserved concessions and privileges should be abolished, including direct and indirect subsidies to companies, no matter what reasons may be offered to justify them.

In his 2001 AFA speech Putin bemoans the fact that over 60 per cent of all industrial investment is directed to the fuel and energy complex, while other sectors of the economy have not been able to attract sufficient levels of investments. He underlines that the accessibility and the development of the transport and energy infrastructure are of exceptional importance for the country. Further, he notes that Russia is poised to reform its power engineering, gas supply, road transport and communications sectors, and calls for greater transparency in money spending. He also warns that cross-subsidies will not be allowed, and calls for sober calculation of the economic and social costs of the transformation of the economy and of Russia's monopolies. Again he notes his displeasure with the fact that the economy still relies preponderantly on rents rather than production, and that the country makes its money mainly from oil, gas, metals and other raw materials. He adds concern that the additional income from export is used for everyday needs, or to feed the flight of capital or, at best, is re-invested in the production of raw materials. Putin further observes that the structure of the economy has not been modernized – indeed, he notes, it is increasingly based on raw materials, making it more dependent on short-term trends.

In his 2002 AFA speech, Putin sees the first signs of improvement: the country has managed to pay off foreign debt in time, or even in advance. He also expresses satisfaction at the fact that exports of non-raw materials have begun to increase, and that, after a ten-year hiatus, Russia now ranks second in the world in volumes of oil production, and first in trade of energy resources. However, he warns that the country must reform its housing and communal services, because only such a reform can get consumers to economize on light and heat, and get producers to use energy-efficient equipment and install devices for recording actual resource consumption.

In the same speech, Putin presents his views on the importance of energy in pursuing foreign policy goals in the post-Soviet space. Here he mentions large joint Russian–CIS projects in infrastructure, transport and energy as important tools in the realization of Russian foreign policy aimed at raising the durability and economic integration of the CIS countries.

Moving on to his 2003 AFA speech, we may note that Putin singles out for special mention the 18 per cent increase in exports of oil, petroleum products and gas. This increase, which has made Russia the largest exporter of fuel and energy resources in the world, he sees as one of the main successes of the previous year.

One year later, in his 2004 AFA speech, Putin shifts the focus to the poor condition and low density of the road network, oil pipelines, the gas transport system and power industry infrastructure as serious structural constraints on the development of the economy. He elaborates further on these issues when discussing various plans aimed at diversification of routes for delivering Russian oil. Specific mention is made of the Baltic Pipeline System (BPS), opening the Western Siberia–Barents Sea pipelines, new routes from oilfields in Eastern Siberia, pipelines bypassing the Bosporus and Dardanelle Straits, and integrating the Druzhba and Adriya oil pipelines.

In his 2005 AFA speech, the Russian president takes up the issue of energy in the context of discussions on regional policy and the planned mergers of several regions. Here he focuses on the unification of Krasnoyarsk krai, Taimyr and Evenk autonomous districts, which, in his opinion, would help the development of new deposits of natural resources and provide the eastern Siberia with uninterrupted energy supplies.

Putin also refers to the development of infrastructure, including energy infrastructure, in other strategically important regions of Russia, such as the Far East, the Kaliningrad region and other border areas. The creation of transcontinental corridors is mentioned as a part of these infrastructural plans.

In his 2006 AFA speech, Putin pays attention to the question of technological backwardness, *inter alia* in the energy sector, and how it results in less efficient use of energy in Russia compared with the country's direct competitors. He calls for Russia to realize its full potential in such high-tech sectors as modern energy technology, transport and communications, space and aircraft building, in order to transform the country into a major exporter of intellectually-based services.

He adds that there is an evident need for more rapid technological modernization in the energy sector, and that Russia needs to develop modern refining and processing facilities, improve its transport capacity and develop new and promising markets. He underlines the importance of meeting Russia's own internal development needs and Russia's external obligations during this process of modernization. In the same speech he also discusses the development of the nuclear energy sector, which is to be based on safe, new-generation reactors and is to help Russia strengthen its position on the world markets for nuclear energy sector technology and equipment. Further, he signals Russia's interest in developing new energy technologies, like hydrogen and thermonuclear energy, and in improving energy efficiency of economy. Here he links this issue not only to technology but also to environmental security and the quality of life for ordinary Russians. He sees implementation of those measures as essential for securing a leading and stable position for Russia on energy markets in the long term, so that the country may play a positive role in forming a common European energy strategy. Putin goes on to note that Russia's planned development of nanotechnology could help it in addressing the issue of energy conservation. In addition, he proposes that markets be organized on Russian territory for trading oil, gas and other goods, with the transactions carried in roubles.

In his 2007 AFA speech, Putin links the need for using more energy-efficient and resource-saving technology to discussions on the development of the housing sector. He praises the idea of establishing the Stabilization Fund but calls also for some changes in its structure and functions. As the funds for the Stabilization Fund have been generated partly by the production and export of oil and gas, these decisions have had an impact on the development of that sector of the Russian economy. Putin proposes the establishment of a Reserve Fund, intended to minimize the risks resulting from the volatility of oil price on global markets. Moreover, part of the oil and gas revenues should go into the federal budget, as a source of financing of large-scale social programmes. The remaining oil and gas revenues are to be placed in the Fund for Future Generations or, as Putin calls it, the National Prosperity Fund.

In the same speech Putin announces the launch of what he refers to as the biggest structural reform project of this sector in decades, which he also describes as the 'second electrification' of Russia. The aim is to increase electricity production by two-thirds by 2020; to this end, both state and private companies are to invest around 12 trillion roubles. The project is to result in the construction of new power plants, including 26 nuclear plants, and will increase the share of nuclear, coal-based and hydroelectricity production. Further, he signals that

Russia will focus more on developing its hydroelectricity production potential by, *inter alia*, constructing new hydroelectric stations in Siberia and the Far East. As Russia has enormous coal reserves as well, Putin also calls for increasing the share of coal in electricity production.

Putin also uses this 2007 address to take up the issue of energy in the context of Russia's relations with its foreign partners. Russia will not act in an arrogant manner – but, he notes, is prepared to defend its economic interests and make use of its competitive advantages, just as all countries around the world do.

Another issue Putin mentions is the need to spend funds generated by the energy sector in order to resolve the housing question. Here he does not mince words, stressing that a country with such massive reserves built up through oil and gas revenues cannot accept that millions of its citizens are living in slums.

Discussing in greater detail the use of natural resources in Russia, Putin underlines that even though Russia was the world's biggest oil producer in 2006, it lagged behind in oil refining. He also expresses concern at the fact that Russian oil fields have flared more than 20 bcm of associated gas every year.

Medvedev's Ideas on Energy: AFA 2008–2011

In the four addresses to the Federal Assembly that Dmitrii Medvedev delivered during his four years at the helm of Russian politics, he paid some attention to energy-related issues. He mentions energy in a proper energy context 28 times, and focuses on oil (8 mentions) and gas (4), as well as energy efficiency (mentioned seven times). We find no mention of either hydropower or coal, and note the scant little attention paid to renewables (industrial waste is mentioned as a source of energy in 2009). Like his mentor, predecessor and successor, Vladimir Putin, when Medvedev makes any reference to climate in these speeches, it is always the investment climate that he has in mind.

In his 2008 AFA speech, Medvedev calls for setting new rules and building new global financial architecture to help to prevent new crises from emerging and to protect the interests of all actors in a more just manner. Such a new architecture, he holds, can play a crucial role in preventing new crises and in minimizing their impact also in the energy sphere. He expresses the view that Russia should become one of the world's leading financial centres, and that rouble should be introduced as the currency for transactions in the oil and gas markets. In other words, Russia should use its energy resources so as to turn its currency into one of the global currencies and Russia itself into one of the global financial centres. This, he says, will mean greater interdependence for all actors involved, resulting in a safer and more sustainable global development.

In the 2009 AFA speech, he focuses on the need to modernize the Russian economy, referring to his own article *Go Russia!* published two months earlier (Medvedev, 2009). Russia's power in the modern world, he states, should be based not on past achievements but on new ideas. He refers to Russia's energy commodities, noting how they generate most of the country's budget revenues, but also mentions that they were developed during the Soviet period. He underlines their importance in helping to keep Russia afloat, but adds that these resources are rapidly depreciating. He calls for transforming Russia from what he himself describes as 'a primitive raw materials economy' into 'a smart economy producing unique knowledge, new goods and technology of use to people'. Later in the speech he lists chronic backwardness, dependence on the export of raw materials, and corruption as the main challenges facing Russia on its way to that brighter future. He expresses regret that Russia has not been able to free itself from what he calls 'a primitive economic structure and humiliating dependence on raw materials', and notes how 'the habit of living off export earnings' is still holding back innovative development. He also accuses the political class of being lulled into believing that oil prices will rise endlessly and that structural reforms can wait. He calls for the quick modernization of Russia – which he describes as a precondition for Russia's very survival in the modern world.

The process of modernization of Russia should, in Medvedev's view, lead to the introduction of the latest energy technology, radically improving energy efficiency. Justifying the focus on energy efficiency with reference to the need to preserve the country's natural resources and pass them on to future generations, he calls upon all Russians to take their share of responsibility for saving energy. Medvedev goes on to state that the government has decided to implement certain measures in order to bring this about, including the production and installation of household-level meters, a transition to energy-saving light bulbs, making city districts more energy-efficient by modernizing utilities networks, introducing a system of payments for services, and making Russia a leader in developing innovation in traditional as well as in alternative energy. He specifically mentions the practice of flaring of associated gas as an example of ineffective use of energy. On the positive side he notes the use of bio-resources – especially timber, peat, and industrial waste – as energy sources. In addition he calls for using modern technology, such as superconductors, to reduce energy loss in transmission systems, and refers to the development of new nuclear technology as a step in the right direction.

One year later, in his 2010 AFA speech, Medvedev presents a plan to improve the energy efficiency of the Russian economy by 40 per cent by 2020. He also proposes directing at least half of the saved budget funds and part of the extra income towards modernization projects, such as increasing energy efficiency, and the development of new technologies in the sectors of energy, IT, telecom and medicine. Further, he calls for building a global or European emergency management system, mentioning his own initiative on protection of the marine environment from oil spills.

In his 2011 AFA speech Medvedev reminds the audience of how he has initiated a shift in Russia's development strategy towards the modernization of the country's economy. He cites the example of reducing the energy intensity of the Russian economy as one of the achievements of his presidency. He also returns to his institutional initiatives from previous years – including his ideas on establishing an international mechanism to prevent and deal with the effects of disasters on the continental shelf, on strengthening the international regime governing nuclear safety, and on the establishment of a new legal framework for energy security. In addition Medvedev lauds the idea of setting aside a part of the oil revenues, a policy that, in his opinion, has helped Russia to prevent a currency crisis.

Dmitrii Medvedev's Address to the Federal Assembly on 22 December 2011 could be rightly described as his political swan song. When Medvedev and Putin had announced on 24 September 2011 that Medvedev would be given the post of prime minister after the elections in which Putin sought re-election as the country's new/old president, this declaration confirmed that Medvedev did not expect to be treated as an independent political player but was, like Igor Sechin, one of the most loyal members of Putin's team.

Sechin's Ideas on Energy

Whereas both Putin and Medvedev have expressed their views on major energy-related issues many times in the public space, it is far more difficult to identify the opinions of the third Russian energy boss, Igor Sechin. Sechin is generally known for his unwillingness to talk to the media, but has, however, shared his views on some issues.

In a lengthy interview given to Reuters in June 2010 (Stott and Faulconbridge, 2010) Sechin takes up various issues. He supports Medvedev's view that Russia should become less dependent on oil and gas, while also stating that the strong role played by the Russian energy sector in the economy is a fact one must live with, because the exploitation of natural resources is, as he puts it, 'the foundation of the Russian economy'. He also warns against hastily selling Russia's energy assets too cheaply, and ensures his interlocutors that, although many assets may have been improperly privatized, the Russian state has no re-nationalization plans. In the same interview Sechin touches on Russia's role as a supplier of gas to Europe. Here he asks Russia's European partners to present more consistent prognoses on future gas demand so that Russia can take proper measures for securing supplies. He also calls on Gazprom to improve its position by improving the efficiency of its operations and diversifying its markets, but adds that the government has no plans of cancelling Gazprom's monopoly on the export of gas from Russia.

In an article published in *Izvestiya* in June 2010, Sechin outlines Russia's reading of the situation as to energy cooperation with its partners (Sechin, 2010). He praises Russian companies for their transparency, and repudiates all accusations of bad intentions in pursuing its goals in the energy sphere. Russia's sole long-term interest in energy cooperation, he stresses, is the creation of a sustainable energy market. He also rejects accusations that Russia is responsible for the high gas prices that customers must pay for energy commodities, pointing out that these prices are formed in several stages and are influenced by the policies of the transit countries and by taxation levels in the final markets. The situation in the global gas markets he describes as unstable: here he notes uncertainties regarding the demand for gas, the development of the LNG market, methane gas and price formation and the many factors that will influence the development of the shale gas market in Europe. In this article, Sechin concludes that the main challenge facing Russia is to modernize its economy, to enable the country to become more competitive in the broader globalization context.

In 2011 Sechin presented his views on energy policy at the Saint Petersburg Economic Forum (Sechin, 2011a), discussing, *inter alia*, the role of Russia on the global energy market. Here he presents Russia as an important energy producer. He expresses the view that Russia will discover new energy resources, and outlines Russian energy priorities. The top priority for Russia in the energy sphere is now to meet domestic energy and budgetary needs. He adds that the Russian leadership expects economic growth in Russia to result in greater demand for energy; and notes that the domestic market had become more attractive to Russian producers in economic terms, in some cases even more attractive than foreign markets. The Russian authorities, he explains, intend to implement certain tax and custom measures that will make it more attractive to process more oil in Russia and export more petroleum products instead of crude oil and gas. Further, Russian energy policy will take into account the needs of domestic as well as foreign energy consumers. He invites both Russian and foreign actors to share the risks connected with the development of new oil- and gas-producing fields: 'consuming countries should also make a contribution' (Interfax, 2011: 16).

Sechin's three main conclusions in his Saint Petersburg Forum presentation (Sechin, 2011b) can be summed up as follows: Russia should diversify its energy markets; Russia should cooperate with the major global energy producers; and Russia should find ways of cooperating with its main energy consumers that can be advantageous for both parties. When discussing market diversification he pays considerable attention to China's growing demands for energy, presenting the highly optimistic IEA data on the Chinese needs for gas imports (150 bcm in 2020 and 330 bcm by 2030). Sechin also mentions the need for building a more stable institutional and legal framework for global and regional trade in energy, and presents the official Russian position on the importance of long-term contracts. Finally, he underlines the importance of interdependence in energy relations with other countries, noting that Russia has become an integral and well-integrated part of the global economy. Energy cooperation between Russia and other countries should be based on principles of equality, mutual interest, and long-term thinking, resulting in greater mutual trust.

What about Actions?

The views of Putin, Medvedev and Sechin on the role of energy in Russia – and on Russia's energy role in the globalized economy – have been heavily influenced by the dramatic events of recent years, with the financial crisis spurring serious debate on the future of the energy sector. Putin seems to have recognized the key challenges, but failed to present a comprehensive approach for solving the problems. Medvedev, whose ideas have obviously been influenced by the experience of the crisis in 2008, managed to identify key issues and present a programme for modernizing the Russian economy that also aimed to deal with the problem of overdependence on revenues generated by the energy sector. However, as it turned out, Medvedev had neither time nor political clout to translate those ideas into political action. Sechin, who had the political responsibility for the workings of the energy system, presented a pragmatic approach: he sought to make the energy system run more smoothly without implementing revolutionary changes, and to involve major energy customers more actively in helping Russia deal with its problems by securing demand and providing the necessary know-how and financial support.

Let us now see how the statements of these three key actors have – or have not – been translated into political action in the sphere of energy in the post-crisis period. By 'action' we understand here the steps that have been taken in order to enable the sector to respond to the serious challenges revealed by the economic crisis of 2008. It took some time before policy makers decided to take those steps, and it is going to take some time before the effects of those actions will be visible in Russian energy sector – but it is important to see how the political class decided to meet the challenges.

There have been several important public discussions on the future of Russian energy sector, followed by a series of moves intended to shape this future. Central here were the meetings of the Russian energy elite in Salekhard in September 2009 and in Novyy Urengoy in October 2010, as well as the discussions in the Security Council of the Russian Federation in December 2010. In addition, an updated version of the Russian energy strategy until 2030 was released at the end of 2009[13] and a geological strategy until 2030 was adopted in June 2010.[14] Also Medvedev's decision to launch an ambitious modernization programme in September 2009, his orders on removing the state officials from the boards of Russian energy companies issued in March 2011, his decision not to run for office in the 2012 presidential elections, and finally Putin's decision to launch an anti-corruption campaign in December 2011 – all these will have mid- and long-term consequences for the realization of Russian energy policy.

The Salekhard meeting was devoted to the future of the Russian gas sector. Prime Minister Putin used the opportunity to invite foreign companies to invest in the sector and to join Russian actors in developing the gas deposits on the Yamal Peninsula: 'We are ready for a wide partnership and this is why we invited you here to Salekhard', he declared, and went on to say that the Russian authorities wanted Western companies to see 'that Russia is working openly and transparently'

---

13 English version available at: www.energystrategy.ru/projects/docs/ES-2030_ (Eng).pdf.

14 See www.vz.ru/economy/2010/6/16/410985.html.

and that 'if something is not clear then you have the chance to clarify it'.[15] Putin explained that Yamal was set to become Russia's new oil and gas province, and that the Yamal reserves would play an important role in stabilizing the world's natural gas markets. Both Putin and Minister of Energy Sergei Shmatko called for the construction of not only new pipelines but also of LNG facilities. The latter would make supplies more flexible and help Russia to diversify its markets.

What made the Salekhard meeting special was the presence of key foreign actors from the gas sector. The top managers of no less than 12 leading foreign oil and gas companies were present: from ConocoPhillips (USA), Eni (Italy), E.ON (Germany), ExxonMobil (USA), GDF SUEZ (France), Kogas (South Korea), Mitsui (Japan), Mitsubishi (Japan), Petronas (Malaysia), Shell (the UK/Netherlands), Statoil (Norway), and Total (France). The fact that these companies were invited was widely seen as a clear sign of Russian interest in attracting investments and gaining access to modern technology for more efficient development of the Yamal reserves.[16]

The meeting in Novyy Urengoy on 11 October 2010 had a slightly different focus, and did not include representatives of potential foreign investors. The main topic here was the future of the entire Russian gas sector, with discussions centring on the 'General Plan for the Development of the Russian Gas Sector until 2030'.[17] This document addresses several key questions influencing the future of the gas sector:

- future demand for gas, domestically and internationally
- production capacity
- environmental issues
- technological aspects (R&D)
- gasification of Russia's regions
- investment needs
- the international market situation
- measures to stimulate the development of the gas sector in Russia.

The main message of the meeting was that Russian gas production was to rise to 1000 bcm per year by 2030, with exports doubling to 455–520 bcm per year. In addition, the government gave the green light for construction of an LNG plant on the Yamal Peninsula; it promised tax breaks to those investing in the sector, and expressed hopes that the share of the 'independent producers' in gas production would increase from the current 20 per cent to 30 per cent by 2030 and the share of new regions (mostly Yamal and Eastern Siberia) from 2 per cent to 40 per cent over the same period. According to official plans, 25,000 km of

---

15   AFP, 24 September 2009. See also http://minenergo.gov.ru/press/doklady/1824.html.

16   http://kommersant.ru/doc.aspx?DocsID=1271371&stamp=634000235263486028.

17   Available at www.energyland.info/files/library/112008/7579b56758481da282dd7 e0a4de05fd1.pdf.

new pipelines and 116 compressor stations would be constructed, requiring investments in the range of 12.3 to 14.7 trillion roubles.

A third meeting took place on 13 December 2010, when the Russian Security Council was convoked to discuss national energy security.[18] The meeting was attended by all members of the Council as well as several invited guests. In his opening remarks, President Medvedev underlined that energy security is crucial to the sustainable and sovereign development of Russia: it influences Russia's social and economic development, its competiveness on international markets and its prestige. Further, the Russian authorities should focus on

- the energy needs of the population and economic actors
- development of hydropower and alternative sources of energy
- quick response to extraordinary situations
- modernization of the energy complex
- protection of its infrastructure against possible terrorist threats
- international cooperation in the sphere of energy.

According to information presented at this meeting, 50 per cent of Russia's oil reserves had already been produced – but even more important was the extensive wear and tear, affecting 60 per cent of the equipment in the gas and electricity sector and almost 80 per cent in the oil refining sector. This means an urgent need for investments and modernization, as well as for a comprehensive approach.

On 30 March 2011, Medvedev announced that representatives of the state were to be removed from the boards of state-owned companies, also those operating in the energy sector.[19] This was among the measures intended to improve the investment climate in Russia, announced during his visit in Magnitogorsk. The removal of state officials from the boards of companies would lessen state influence on strategic decisions made by those bodies and could be seen as a step in the implementation of the liberalization and modernization programme announced in Medvedev's *Go Russia!* article, published in September 2009 (Medvedev, 2009). Medvedev's announcement on 30 March 2011 was meant as a signal to potential foreign investors, encouraging them to invest in Russia – and its energy sector – where their money and know-how were needed to enable the continued provision of oil and gas to Russian and foreign markets. Medvedev's decision was intended to improve Russia's image and the framework conditions for operations in the Russian economy and the energy sector.

State officials were required to vacate their posts by 1 October 2011. In particular, this meant that Igor Sechin would have to leave his posts in Rosneft, Rosneftegaz and in INTER RAO UES; and Energy Minister S. Shmatko would leave the boards of RusGidro, Gazprom, Zarubezhneft.[20] In fact, both men

---

18 For further information, see www.kremlin.ru/news/9809 and www.vz.ru/news/2010/12/13/454450.html.

19 http://lenta.ru/news/2011/03/30/medvedev/.

20 http://news.kremlin.ru/ref_notes/900.

did vacate their posts in those major energy companies by the deadline[21] – but another intrigue was underway. It was also expected that the First Deputy Prime Minister V. Zubkov would leave his post in Gazprom, although this was not on the official list of the posts to be vacated. By the end of September 2011, however, it had become clear that Zubkov would retain his position in Gazprom.[22]

In the meantime, on 24 September 2011 it had been announced that it was Putin and not Medvedev who would run for presidency in 2012 – a decision with huge implications for the future of not only the energy sector but also of the Russian state project. Putin's anti-corruption campaign, launched in December 2011 and directed against those who channelled revenues and other assets to offshore tax paradises, was widely seen as preparing a basis for his presidential campaign. This move resulted in the forced departure of many managers of state-owned energy companies. The campaign should be seen not only as a propaganda move for boosting Putin's ratings on the eve of elections, but also as an effort to tackle one of the most important challenges facing those interested in investing – the issue of corruption and lack of transparency (Koroleva, 2011; Ponomariev, 2011; Yakovenko, 2012).

However, it is difficult to gauge how the earlier actions of Putin and Medvedev influenced several major Western energy actors who decided in 2010 and 2011 to increase their presence in Russia. Some of these deals – like the one between Rosneft and Chevron announced in June 2010, or between BP and Rosneft announced in January 2011– ended in failure, for various reasons. But there are others – like the Rosneft–Exxon deal first announced in January 2011 and then extended in August that year, or the Novatek–Total deal announced on 2 March 2011 – that seem more promising, and might result in closer long-term cooperation between Russian companies and their Western partners.

But how did Putin's and Medvedev's actions shape the conditions for Russian and non-Russian actors operating in the country's energy sector after the March 2012 presidential elections and the swap of functions between the two? Putin's formal return to the presidency in May 2012 might have secured continuity and a certain level of stability and predictability, but not necessarily better framework conditions for Russia's economy in general and its energy sector in particular. Both Putin and Medvedev have signalled repeatedly that Russia needs a boost, and that the energy sector alone cannot solve all Russia's problems – and is itself in need of deep structural reforms. How Putin will cope with those serious challenges as the country's new old president will to a large extent define his historical role in Russia.

After winning the elections in March 2012 and re-assuming presidential power in May 2012, Putin decided to reverse some of the decisions taken by Medvedev and his team. The most crucial development was the strengthening of the role of Igor Sechin as the key actor in Russian energy sector. Although he had been fired

---

21   http://interfax.ru/politics/txt.asp?id=210250.
22   http://lenta.ru/news/2011/09/25/zubkov/.

from Medvedev's government, his role in organizing the work of the country's energy sector was strengthened when he was in 2012 appointed not only CEO of Rosneft, the main state asset in the oil sector, but also the secretary to the Presidential Commission for Strategic Development of the Fuel and Energy Sector and Environmental Security created by Putin in June 2012.[23] That strategic move has had several consequences. On the one hand, it restricted the power of the Russian government, in which Arkadiy Dvorkovich was given formal responsibility for managing energy-related questions, to shape the future of the sector. On the other hand, Sechin managed to realize his – and Putin's – pet project, the restructuring of the Russian oil sector, that was achieved by building Rosneft's strategic alliance with the BP and 'dissolving' TNK-BP.[24]

The decisions of Putin and his team are certain to have a huge impact on developments in the Russian energy sector and on its ability to cope with changing market conditions. Many of those issues are discussed an analysed in the ensuing chapters of this book. In the final chapter, we return to those core questions, and offer some conclusions on the possible effects of Putin's policy.

References

Balzer, H. 2005. The Putin thesis and Russian energy policy. *Post-Soviet Affairs*, 21(3), 210–25.

Balzer, H. 2006. Vladimir Putin's academic writings and Russian natural resource policy. *Problems of Post-Communism*, 55(1), 48–54.

Godzimirski, J. M. and E. Wilson Rowe 2008. Developments in energy discourses in the early days of the Putin–Medvedev tandem. RUSSCASP Working Paper. Available at: http://www.fni.no/russcasp/Energy_in_Russian_Politics_russcasp_working_paper.pdf [accessed: 5 December 2012].

Goldsworthy, B. and D. Zakharova. 2010. Evaluation of the oil fiscal regime in Russia and proposals for reform. IMF Working Paper 10/33. Available at: http://www.imf.org/external/pubs/ft/wp/2010/wp1033.pdf [accessed: 5 December 2012].

Interfax 2011. *Russia & CIS Oil and Gas Weekly*, 16–22 June 2011.

Koroleva, A. 2011. Goskompanyam prigrozili [State agencies were warned]. *Expert* 20. Available at: http://expert.ru/2011/12/20/goskompaniyam-prigrozili/ [accessed: 5 December 2012].

Kryshtanovskaya, O. 2011. The tandem and the crisis. *Journal of Communist Studies and Transition Politics*, 27(3/4), 407–19.

Medvedev, D. 2009. *Go Russia!* Moscow: Kremlin. Available at: http://archive.kremlin.ru/eng/speeches/2009/09/10/1534_type104017_221527.shtml [accessed: 10 November 2012].

---

23 http://eng.state.kremlin.ru/commission/29/news/4023.
24 http://lenta.ru/articles/2012/10/22/rsnft/.

Orlov, D. 2012. *100 veduschikh politikov Rossii v 2011 godu [100 leading politicians in Russia in 2011]*. [*Nezavisimaya Gazeta* online]. Available at: http://www.ng.ru/ideas/2012-01-16/9_top100_2011.html [accessed: 15 November 2012].

Ponomariev, V. 2011. Pilotnaya zachistka [Pilot cleaning]. *Expert*, 27 December. Available at: http://expert.ru/2011/12/27/pilotnaya-zachistka/ [accessed: 15 November 2012].

Pribylovskiy, V. 2010. *Vlast-2010. 60 biografiy*. Moscow: Tsentr Panorama.

Putin, V. 1999. Russia at the turn of the millennium, published on 30 December. Available originally at: http://www.gov.ru/ministry/isp-vlast47.html [accessed: 15 November 2012].

Putin, V. 2012. *Itogi desyatiletiya i predstoyaschie vyzovy [Results of the 10 years and future challenges]*. Available at: http://www.putin2012.ru/program/1 [accessed: 5 December 2012].

Reznik, I. and I. Mokrousova 2012. Igor Sechin, pervyy vozle Vladimira Putina [Igor Sechin near Vladimir Putin]. *Vedomosti*, 19 March. Available at: http://www.vedomosti.ru/library/library-investigation/news/1541119/pervyj_vozle_putina [accessed: 5 December 2012].

Sakwa, R. 2010. *The Crisis of Russian Democracy. The Dual State, Factionalism and the Medvedev Succession*. Cambridge: Cambridge University Press.

Sakwa, R. 2011. Russia's grey cardinal [online: Post Soviet World]. Available at: http://www.opendemocracy.net/od-russia/richard-sakwa/russias-grey-cardinal [accessed: 15 November 2012].

Sechin, I. 2010. Skazki pro gaz i pro nas [Fairy tales about gas and us]. *Izvestiya*, 29 June. Available at: http://www.izvestia.ru/economic/article3143376 [accessed: 15 November 2012].

Sechin, I. 2011a. *Novye puti dostizheniya energeticheskoj bezopasnosti St Petersburg [New ways of achieving energy security, Saint Petersburg]* [online: Forum SPB]. Available at: http://www.forumspb.com/userfiles/files/Presentation_RUS.PDF [accessed: 15 November 2012].

Sechin, I. 2011b. Sechin's comments in discussion on new ways of addressing energy security. *2011 Sankt Petersburg Economic Forum St Petersburg*. Available at: http://www.forumspb.com/upload/shorthand/shorthand_161_ru.pdf.

Stott, M. and G. Faulconbridge. 2010. *Putin's right-hand man exits Kremlin shadows* [online: Reuters]. Available at: http://www.reuters.com/article/2010/06/20/us-russia-sechin-idUSTRE65J0G120100620 [accessed: 15 November 2012].

Ukhov, V. 2008. *Rasklad v verkhakh* [online: New Times]. Available at: http://shop.newtimes.ru/articles/detail/4601 [accessed: 15 November 2012].

Vedomosti. 2008. *Smena protivnika* [online: Vedomosti]. Available at: http://www.vedomosti.ru/newspaper/article/2008/03/06/143090 [accessed: 15 November 2012].

Yakovenko, D. 2012. Bolshaya proverka [Big test]. *Expert* [online], 4 (787). Available at: http://expert.ru/expert/2012/04/bolshaya-proverka/ [accessed: 15 November 2012].

Chapter 2
Oil Industry Structure and Developments in the Resource Base: Increasing Contradictions?

Valeriy Kryukov and Arild Moe

Introduction

Russia is a major international energy player but petroleum is also crucial to Russia's economy, providing perhaps as much as 40 per cent of GDP, and 60 per cent of export earnings. Whereas natural gas often attracts most attention in analyses of Russia's energy power due to the physical imprint of supplies and the dependence reflected in pipelines, oil remains of greater importance for the Russian economy. That is why this chapter focuses on developments in the oil sector.

As in many other issue areas, the sheer size of Russia must be taken into consideration when the situation in the oil sector is analysed. Although the energy resources are enormous, they are spread out over large territories, some of which are very poorly developed in infrastructure. This obviously has consequences for time and costs in developing the resources. The general trend has been movement from west to the east, from south to the north, from onshore to offshore, and from relatively shallow to deeper geological strata. Also the composition of the reserves to be developed is changing, from relatively simple and well-known compositions like light oil, to more complex ones.

The Russian hydrocarbon resource base is not being exhausted in absolute terms. At the end of 2011 the reserve to production rate was 23.5 years, according to BP Statistical Review of World Energy 2012 – a higher rate than in many other oil-producing countries. This corresponds to some 10.5 billion tons. Russia's own figures on oil reserves are in principle kept secret, but it seems that they are significantly higher, some 22 billion tons (Shmatko, 2010). There is, however, much disagreement about how well these figures reflect the real

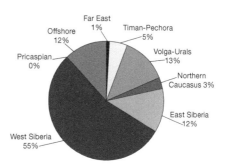

Figure 2.1   Remaining oil resources
Source: Bushuev and Gromov, 2007.

production potential (see below), and it cannot be denied that we are witnessing the exhaustion of the best located and relatively uncomplicated deposits.

This chapter reviews developments in the resource base for oil, discussing whether the structure of the Russian oil industry and policy trends in this regard fit the resource situation. The underlying argument is that the future of Russian oil production will depend on alignment of the resource picture, industry structure and policies.

Resources and Reserves – Location, Size and Quality

By 2011, some 1640 oil fields were in production in Russia; and 62 per cent of the reserves in these fields had already been produced. The situation varies across the country. Most of European Russia (Tatarstan, Bashkortostan, Komi) is characterized by high depletion levels. But also the mainstay of production – West Siberia – is becoming depleted. Average production in Russian fields is falling, to 11 tons per day. The water content is 85 per cent, due to water injection to boost output (Prischepa and Podol'skiy, 2012). Somewhat better is the situation in new oil areas: these include the north of European Russia (Nenets Autonomous District) and East Siberia (Krasnoyarsk, Sakha-Yakutia), as well as the Arctic continental shelf and the Far East.

In the 1990s and well into the new century, replenishment of reserves was an increasing concern. In the period 1994 to 2004 less reserves were added than what was being produced. More recently, this problem appears to have been solved, with additions to reserves exceeding production by a considerable margin (see Figure 2.2).

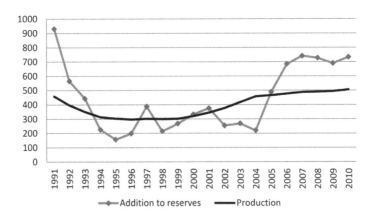

Figure 2.2   Reserve additions vs. production (mill. tons)
Sources: Sadovnik, 2006; Orlov, 2008; *Rossiyskie nedra*, 4, 2009; *Neftegazovaya vertikal'*, 8, 2012.

But what does Figure 2.2 really tell us? Intrinsic to the mining industries is the level of exploration and the various reserve and resource categories, basic issues often overlooked in broader political overviews of resource politics (Kryukov and Moe, 2007). The reserve classification system used in Russia operates with categories A, B and C1 for explored reserves, representing different levels of geological certainty and accuracy. Categories A and B are reserves already in production, whereas C1 are reserves that have been discovered and delineated but not yet exploited. Altogether these three categories are often translated as 'proven reserves', the term used in most Western countries.

Then there is the category C2. It is more ambiguous, and is sometimes translated as preliminarily estimated reserves, estimated only after geological and geophysical work, not drilling, in the immediate vicinity of producing fields. Categories C3, D1 and D2 represent unproven *resources* with high/very high degrees of uncertainty.

For a balanced resource management policy it is essential to 'move' resources from the uncertain categories into the more certain ones, by geological exploration in various categories of resources. Thus some resources in category D2 will be moved into D1, some resources from D1 will become C3, and some from C3 will become C2. Finally, some resources in the C2 category will become explored reserves, and ultimately ready for exploitation. This latter 'movement' is the additions to reserves reported in Figure 2.2.

In most countries it is the responsibility of the state itself to carry out basic seismic surveying and thus identify resources for further exploration. The oil industry will be preoccupied with the higher levels of the 'reserve pyramid': bringing resources into the C2 category and further into the higher reserve categories. And here lies a major problem. Whereas the Russian oil industry previously took it upon itself to 'prove up' resources from the C3 to the C2 category, the industry has for many years now been concerned only with exploration in the vicinity of ongoing production – bringing resources from C2 to C1. Such work is far easier to carry out, and entails a much higher return on investments than is the case with exploration in the less certain resource categories. In addition, reserves in fields adjacent to ongoing production areas are almost automatically added to the reserves of the company already operating in the area, whereas it is not certain that a company that has carried out successful exploration in a virgin territory is going to be granted a production license.

The behaviour of the oil companies is rational and understandable from a commercial point of view, prioritizing relatively low risk and relatively fast payback in an uncertain institutional environment.

And the focus on the relatively easy accessible C2 category has, together with the even larger contribution from reappraisals of already discovered fields (see below) helped to compensate production with new reserves, as shown in Figure 2.2. But this development tends to overshadow problems further down in the 'reserve pyramid', which should be the concern of the state's resource management policy. Little goes on between the C3 and C2 categories. Oil companies show scant interest in applying for licenses to explore resources in the C3 category. In 2011, only 25 per cent of the oil and gas licenses put up for auction found takers (*Neft' i Kapital*, 1–2, 2012). Potentials in the C2 category are likely to come to an end soon, and it will take time before C3 resources can be proven and transferred to the reserve categories, even if policies and incentives were changed today. The whole cycle – from the identification of potential areas, to the discovery of fields and preparing them for exploitation – takes some ten to 20 years. This is a big challenge facing Russian oil production.

But also the quality of the reserves – not only resource figures – is an issue. The Russian reserve classification system tells us what is *technically* recoverable with current technologies. It does not say anything about what is *commercially* recoverable. In Western classification systems costs vs. price is a crucial parameter, which means that reserve assessments will change with changes in the oil price. In Russia that is not the case. In the Russian system it is perfectly possible to operate with reserves that are uneconomical to produce, which would be a contradiction in terms in Western classification systems. The logical suspicion is that Russian figures exaggerate the production potential – but by how much?

C1 is easily the largest of the Russian reserve categories – they were estimated as some 73.5 per cent of reserves in 2000. It has long been argued that this category is too heterogeneous to make much sense, containing reserves that will be very hard to recover. Consequently, lumping reserves in this category together with A and B reserves overstates the true reserve base (Dienes, 2004; Khalimov, 2003). Some estimates indicate that as much as 56 per cent of Russia's oil reserves now fall in the category 'complicated reserves' (Maksimov, 2011): uneconomical to produce under existing conditions. Official figures are lower, but the negative tendency is nevertheless clear (Shmatko, 2010) (see also Figure 2.3).

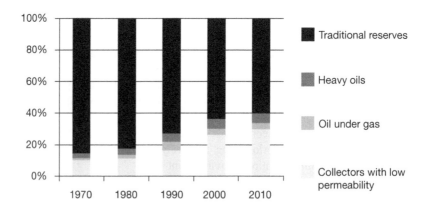

Figure 2.3   Types of oil reserves in Russia 1970–2010
Source: Based on Vygon, 2010.

For many years, attempts have been underway to revise the reserve classification system, basically by giving more weight to economic parameters (see Melnikov, 2012). Implementation of a new system has, however, been postponed several times. We believe a fundamental problem is how to arrive at 'objective' economic indicators, a problem also encountered in taxation reform (see below). In the absence of a market for the various inputs to oil exploration and production, achieving a classification system fully compatible with Western systems is impossible, but all the major Russian oil companies employ various methodologies to simulate western-style estimates. The problem is largest for the authorities who are presented with skewed estimates, and for small companies who do not have the possibility to carry out their own assessments and must rely on the figures based on the outdated methodology.

In most of Russia's oil-producing regions, no major oil discoveries have been made for many years. One reason is that most exploration has focused on areas close to producing fields, revealing only satellite fields that are medium-sized or small. West Siberia has by far the largest explored reserves and will continue to play a dominant role in Russia's oil production for many years. It also is the region with the largest expected undiscovered resources. But, according to some accounts, about 50 per cent of new discoveries are expected to be in deep and complicated layers, and will be too costly to develop under existing conditions, even if they happen to be near existing infrastructure (Prischepa and Podol'skiy, 2012). Much of this is shale oil, or tight oil. Due to the abundance of conventional reserves and low incentives for developing the required technology, these resources have largely been considered uninteresting until now.

The areas where large discoveries can be expected are located far from consumption centres and with sparse or no infrastructure. Both factors contribute

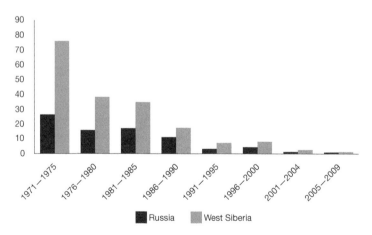

Figure 2.4   Average size of new discoveries (mill. tons)
Source: Neftegazovaya Vertikal', 19, 2010.

significantly to costs. According to estimates from Rosneft, if current production in West Siberia costs approximately USD 30 per ton, the cost at Rosneft's Vankor field in the north of Krasnoyarsk district is about USD 80, including infrastructure costs; at Sakhalin-1, between USD 200 to 300, and Arctic offshore fields are expected to cost as much as USD 500 to 700 per ton to produce (Bokserman, 2011). And, as noted, even if there are ample resources in the ground, it will take many years to develop the necessary infrastructure and bring fields on-stream in these areas.

Are There More Reserves in Producing Fields?

There is considerable agreement on the state of the resource base as depicted above, but more controversy regarding the prospects in currently producing fields. In recent years the average production from each producing oil well has been falling: it was reported to be 7.8 tons per day by 2007 and is prognosticated to continue to fall, reaching 5.4 tons/day by 2030 (Maksimov, 2011). New wells in the northern part of European Russia (Timan-Pechora) and East Siberia are producing some 25 to 30 tons/day, a much lower yield than new wells had in West Siberia (Maksimov, 2011). This development has very much to do with the smaller size of the fields.

A perhaps even more alarming tendency is that average recovery rates have decreased significantly (Agranovich, 2008). The trend is especially worrying in Khanty-Mansiysk Autonomous District where approximately 50 per cent of Russian oil production takes place (see Figure 2.5). More than two-thirds of the resources in fields under production remain in the ground. That is a high figure, also compared with other mature petroleum countries. In the United States, the average rate of recovery increased from 0.29 in 1965 to 0.43 by 2009 (Muslimov, 2009). A major reason for the low rate of recovery in Russia is 'objective', as Russian fields are more complicated. But the rate is also dependent on development strategies as well as incentives – a point to which we return later.

Approximately half the reported addition to reserves since 2000 decade has been accomplished by revising the projected rate of recovery in many fields. This can happen by, for example, implementing new technologies that can secure a higher recovery rate. The question is if these revisions will prove to be realistic.

Investments in drilling, both exploration and production drilling, have increased considerably. This has been portrayed as an important response to the challenges in the resource base discussed above. But what is the net result of the increased financial effort? Table 2.1 shows many of the important trends in the Russian oil sector.

The situation with regard to exploration drilling is problematic. Despite the increased funding, the amount of drilling is going down. The cost of drilling has multiplied. Moreover, the number of discoveries is going down.

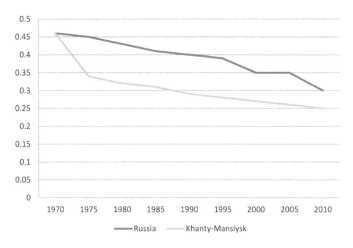

Figure 2.5 Developments in the rate of recovery
Source: Adapted from Muslimov 2012.

Table 2.1 Production and exploration drilling in Russia 2001–2009

	2001	2002	2003	2004	2005	2006	2007	2008	2009
Production drilling (1000 m)	8960	8302	8576	8413	9174	11583	13761	14603	14090
Exploration drilling (1000 m)	1260	766	681	585	635	722	871	852	464
New fields	68	45	31	29	26	26	40	33	26
New wells	6642	3764	3606	3529	3805	4432	5201	5593	5584

Sources: Rosstat; Central dispatching board of the fuel and energy complex.

With production drilling, however, things seem different: more financing has yielded more drilling. But, since oil production has remained more or less stable, the productivity of the production drilling is clearly declining. The number of new wells is fairly stable, but average production from each producing oil well has been falling.

Response and Outlook

The official predictions are that Russian oil production will be maintained at the current level at least until 2020 (Government of the Russian Federation, 2009). What happens thereafter will depend on decisions taken in the coming years – and those decisions will depend on the answers to two fundamental questions: How interested will companies be in increasing exploration of new petroleum deposits? How interested will petroleum-producing companies be in stepping up recovery from existing fields? The answers are by no means given.

Clearly, companies have good commercial reasons for reluctance to take on new mega-projects with high risks. When the Russian government has occasionally criticized the companies for inactivity, the response from industry has been to argue and lobby for concessions, in the form of tax holidays or state-financed infrastructure. As of mid-2011, zero export tax on oil was effective in 21 fields in East Siberia and the Caspian Sea; and zero production tax had been introduced in Russia's Arctic regions (zero tax until accumulated production had reached 25 mill. tons, over maximum 15 years) (Mazneva, 2011). Rosneft's frontier Vankor project in the north of Krasnoyarsk kray reached this level in August 2011, and consequently lost its tax holiday: the state could now tax income from the project. But since Russian taxation is based on production volumes, not on net revenues, the profitability of the project decreased considerably. This tax system gives disincentives to the rational management of resources in the fields. A company will be inclined not to extract as much oil from the field as it could, simply because the marginal cost will be higher than the income it can derive, when taxes are included.

Oil companies strive to reduce the projected production level from a field compared to the level set by the Central Commission for Development of Mineral Fields (TsKR), which is under Rosnedra – the Federal Subsoil Resources Management Agency. Given the taxation system they have good reason for this – moreover, they have the possibility to do so through negotiations in collegial commissions set up to fix and adjust technical development plans for fields (Popov, 2012). Huge recovery losses follow when the oil companies get acceptance for their estimates. By 2010 the discrepancy between actual production and possible production level amounted to 35 mill. tons in West Siberia (Muslimov, 2012).

Clearly the taxation system is a major problem, deterring investments in new production as well as maximum exploitation of existing fields. Oil production is taxed uniformly (with some exceptions) on the basis of volume, not net income. The taxation system in the petroleum sector is not geared towards differentiation. No consideration is given to the specifics of hydrocarbon exploitation – deteriorating production conditions in existing fields and higher costs in new fields. The system is almost exclusively fiscal in orientation. The goal is to maximize state revenues.

If the weaknesses and dysfunctions are so obvious, why, then, has the taxation system not been revamped? There are several reasons. First it should be acknowledged that the system also has a very positive feature: it is simple and easy to administer. This is an important reason why there has been resistance to changing it. However, in principle the authorities have acknowledged that the current system is inadequate, and the introduction of a profit tax to replace the mineral extraction tax has been discussed several times. This thinking gained new momentum with the changes in government from May 2012. The new energy minister Aleksandr Novak has acknowledged that the existing taxation system 'makes extraction of significant reserves in new as well as producing fields unprofitable' (RBK Daily, 14 June 2012). However, opponents of a profits tax argue that it would be too complicated to administer. The tax rate would have to be set individually for every licensed block; and, without unified norms and assessments of plans for development and exploitation of fields as well as real, functioning markets for supplies and services, tax rates would be subject to negotiations and manipulation to an even greater extent than today.

The fiscal concerns remain strong. According to Russia's new energy minister, the state's share of the oil rent is the highest in the world: 56 USD per barrel produced in old fields and USD 37 in new fields (Novak, 2012). Tinkering with the taxation system involves the risk of a shortfall in state revenues, at least in the short term – a very risky proposition. But the opposite position, of retaining the existing system, has its own risk: falling production – with corresponding loss of revenue.

Some tax concessions have been introduced. They ease the tax burden in specific, but vast, geographical regions – notably East Siberia and Arctic offshore – and do not discriminate between projects, fields or wells. Obviously, conditions vary considerably throughout East Siberia, for example, although the average situation is much more challenging than in West Siberia. But a reform that would encompass the bulk of the 1600 fields or so under production is still lacking. 'Testing' a new profit tax system in individual fields is under consideration, however (RBK Daily, 14 June 2012).

Low efficiency and lack of interest in new development is also caused by other framework factors, in particular the inadequate infrastructure in new regions, especially trunk pipelines. If pipeline capacity is insufficient it is impossible to get all the crude oil that Rosnedra maintains is to be produced from a given field to the market. But pipeline development is the remit of the state pipeline monopoly Transneft, and its development plans are not always well coordinated with field developments. Lack of access to transport infrastructure can make it impossible to develop a project, even if an independent company should obtain a license to develop a promising deposit. Reforms in 2011, ostensibly meant to secure non-discriminatory access to trunk oil pipelines, did not provide the flexibility needed for independent producers. The new regulations merely made the pipeline monopoly simpler and more transparent to administer. Access to the pipelines can only be granted on the basis of contracts signed a year in advance; transport fees must be paid whether oil is supplied or not; and transport rights cannot be transferred to other producers. As with the taxation system, the emphasis is on governability and control, even if this creates barriers to effective resource management. Even if such institutional problems or shortcomings are acknowledged, more official attention is devoted to making leaps into new regions that promise new and large discoveries. The Arctic continental shelf and East Siberia play central roles in this regard. How do they fit into Russian policies and institutions?

The Arctic Continental Shelf as a New Energy Province?

The Soviet Union started seismic surveys in the Barents Sea back in the late 1970s, and exploration drilling from the early 1980s. At the time this was widely interpreted abroad as an 'offshore offensive'. In retrospect it seems clear, however, that the main purpose was mapping of resources for the longer term, and that there was no strong drive to start offshore petroleum production (Moe, 2010). Today, Russian estimates put initial hydrocarbon resources on the Russian continental shelf at some 70 billion tons of oil equivalents of hydrocarbons.[1] The lure of the offshore resources lies not only in the big total figures, but also that there are expectations of big fields: '... new Samotlors, new Romashkinos, new Shtokmans ...' (Khramov, 2012). This could make a rapid increase in offshore output possible.

The country's main offshore resources are located in the Arctic, 70 per cent in the Barents and Kara Seas alone. But there are expectations of deposits further east, both in the Laptev Sea and the East Siberian Sea (see Table 2.2). In the latter areas, which are even more remote from infrastructure, the geological mapping has been very limited though, and official Russian estimates of the resource base have not changed much over the last ten years or so. Of the official total (initially) estimated Russian offshore resources, only some 10 per cent, or 7 billion tons, have actually been discovered (Khramov, 2012).

Only when concerns about the onshore resource base – as discussed above – became loud did offshore petroleum production emerge as an important element in Russian energy planning. Starting in the first years of the new century, offshore resources were highlighted in political speeches and declarations, and ambitious production goals for 2020 or 2030 were presented. Government agencies followed up with strategies for Arctic and offshore development (Moe and Wilson Rowe, 2009). There were also other important developments in Russian energy policy. Assets were transferred to the state-dominated companies and they received privileged access to resources, notably through changes in legislation in 2008 that granted Rosneft and Gazprom a monopoly on operating new offshore projects. The two were also given licenses for large areas. However, these two companies did not show the expected resolve offshore. This was understandable, given their limited offshore experience and

Table 2.2   Initial hydrocarbon resources on the Russian continental shelf (billion tons oil equivalents)

Barents Sea	17.9
Pechora Sea	3.6
Kara Sea without gulfs and bays	26.2
Kara Sea - gulfs and bays	4.9
Laptev Sea	2.3
East Siberian Sea	4
Chuchi Sea	0.8
Okhotsk Sea	6.2
Other	4.1

Source: Calculated on the basis of Pavlenko (2011) and Khramov (2012).

---

1   This is a widely reported figure, see for example 'Russian continental shelf', Gazprom   website   http://www.gazprom.com/about/production/projects/deposits/shelf/. Russian reserves and resources are reported in standard fuel equivalents (coal equivalents). 1 ton standard fuel = 0.7 ton oil equivalent. Figures here have been converted from Russian sources according to this rate.

their extensive activities onshore. Here we should note that other companies with offshore experience, notably Lukoil, were barred from operating or seeking licenses.

The discrepancy between official goals and the reluctance of the companies became the target of critical remarks, particularly from the Ministry of Natural Resources. The minister and his deputies started arguing for a change in the laws, so as to allow private Russian and foreign companies to play a role on the Russian continental shelf (Donskoy, 2009). They reasoned that capital needs were very great; state-of-the-art technology would be required, and it was important to spread the risks.

The Arctic offshore impasse was lifted in 2011 when Rosneft signed a 'Heads of Agreement' with BP involving exploration and possible development of a large area in the northern part of the Kara Sea. The deal collapsed due to conflict between BP and the Russian shareholders in the TNK-BP consortium, but the foreign role in the Kara Sea project was taken over by ExxonMobil in a new agreement that was finalized in April 2012 (see Overland et al., 2012). In parallel, after the delimitation agreement with Norway in the Barents Sea entered into force, Rosneft was granted exploration and development licenses for most of the Russian part of the previously disputed area. In April 2012 the company signed an agreement with the Italian company ENI to explore and subsequently develop resources in the southern part of the area, and a few weeks later a similar agreement was made with Norway's Statoil for the northern part and in three blocks in the Okhotsk Sea. Thus within a very short time-span, several major Arctic offshore development plans were presented.

According to Rosneft the Arctic areas to be jointly explored with the foreign companies contain vast resources:[2] with ExxonMobil: 6.268 bill. tons (46 bill. barrels) of oil and 14.58 trillion cubic meters 'estimated recoverable resources';[3] with Eni: 3.5 bill. tons (25.7 bill. barrels) 'recoverable resources'; with Statoil: 2 bill. tons (17.7 bill. barrels) of oil and 1.8 trillion cubic meters of gas 'prognosticated resources'.[4]

Such figures must, however, be treated with great caution. There has not been any drilling in these areas, and assessments are based on fairly superficial geological information. It will take time before the resource picture becomes more certain. More specifically, the agreement with ExxonMobil stipulates seismic surveys in the period 2012 to 2016 and drilling of the first exploration well in 2014 or 2015. The agreement with ENI stipulates first exploratory drilling before 2020,[5] as does the deal with Statoil.[6]

---

2   http://www.rosneft.ru/.

3   When BP entered into a deal with Rosneft over the same blocks in early 2011 it was announced that they were estimated to contain 5 bill. tons of oil and 10 TCM of natural gas. http://www.zerich.ru/news/prime-tass/fr/99126/.

4   Includes north Barents Sea and Okhotsk Sea blocks.

5   http://www.rosneft.ru.

6   http://www.statoil.com/no/NewsAndMedia/News/2012/Pages/StatoilRosneftMay2012.aspx.

Rosneft's deals with the three foreign companies involve similar schemes. Joint ventures (JVs) will be established for exploration of and eventual production from the assigned blocks. The foreign companies will hold one third of the shares in each JV and Rosneft two thirds. The foreign companies will cover all the costs in the geological prospecting phase (seismics) and a certain number of exploration wells, and are also to compensate Rosneft for 'historical costs' – initial prospecting – and a third of the price of acquiring the licenses. At present it seems improbable that the JVs will be allowed to take over the licenses.

Evidently, the foreign companies are expected to take all the risks in the initial exploration phase, even though they are only minority partners in the JVs (Important decisions in the JVs will reportedly have to be made unanimously.) In any case, the foreign partners demonstrate considerable faith in their Russian partner as well as in the framework conditions. These conditions, especially tax rules, have been a matter of concern for foreign companies – and also the Russian companies – interested in Arctic ventures. And the spate of contract-signing in April 2012 was clearly unleashed by Putin's declaration of tax concessions. New offshore projects would be exempted from the export tax (USD 460 per ton as of April 2012); moreover, the minerals extraction tax – NDPI – can be lowered to 5 per cent for the most complicated projects. Property tax and value added tax on imported equipment that Russia does not produce would be lifted. It was promised that the tax conditions would not change for a period of 15 years after start-up of production.[7]

The deals concluded by Rosneft in the course of less than one year mark a significant shift in Russia's Arctic offshore policy, and help realign realities with the ambitious rhetoric heard since the early 2000s. The authorities acknowledge that Russian companies are not in a position – financially and not least as regards technology – to play a dominant role in offshore development, and that partnerships with international companies are required. Moreover, they recognize that the terms must be adjusted if the foreign companies are to commit large funds for investment. But Russia is not ceding control over developments on its continental shelf.

According to Rosneft's president Igor Sechin, production from the blocks developed with ExxonMobil may start in 2020 if all goes well (oilcapital.ru, 19 April 2012). This sounds very optimistic. In addition to the geological uncertainties, the pace of development in other complicated offshore projects internationally would indicate a later start-up. Questions must also be raised regarding Rosneft's capacity to handle all the projects it now will be involved in – if it intends to be at the helm, as presupposed in Russian policy.

The Rosneft deals pre-empted the attempts at liberalizing access to the Russian continental shelf. Conspicuously, they were signed on the eve of the end of Putin government, when Igor Sechin was still deputy prime minister in charge of energy policy. With Putin again becoming president, it might have been assumed that Rosneft's plans would be accepted without much further ado. But the rush to sign

---

7   12 April 2012. 'Prime Minister Vladimir Putin holds a meeting on promoting the development of the continental shelf'. http://www.premier.gov.ru/events/news/18680/print/.

the deals could perhaps reflect fears that the new government might deliberately complicate or delay the process. When the new government headed by Dmitry Medvedev was established, it gave overt support to liberalization offshore (Barsukov et al., 2012), but by then most of the promising offshore acreage had already been given to Rosneft. And Rosneft wants more, in early 2013 the company received 12 new licenses on the continental shelf – which gives it control of about 75 per cent of the most promising area on Russia's Arctic continental shelf (*Russia & CIS Oil and Gas Weekly*. No 42, 2012). Seven of the new licenses were included in an extended cooperation agreement with ExxonMobil.[8]

Not only the licenses for promising areas given to Rosneft, also the alliances formed with foreign companies have limited the scope for potential new Russian entrants, like Lukoil, to the Russian continental shelf. Rosneft could no longer be accused of inactivity, and any attempt to change the allocation of licenses would meet resistance from the international partners. Ironically, having strengthened itself by establishing its own alliances with foreign companies, Rosneft, together with Gazprom, lobbies openly against the idea that private Russian companies could play independent roles on the Russian continental shelf. They are welcome, but only as junior partners to the state-controlled companies, a role they have so far declined (RBK Daily, 10 October 2012).

The Ministry of Natural Resources still harbours ideas of opening up the continental shelf for other Russian companies. And in 2012 the Ministry of Energy, unlike earlier, seemed to be supporting access not only for private Russian companies, but for foreign companies as well (Ft.com, 4 October 2012). Again, as in the case with taxation reform, what is emerging is a sharper discussion over policy direction – and also here the trade-off between control versus efficiency is at the centre of the dispute.

Despite the major break-through in 2012 it is too early to tell whether Arctic offshore resources will become a major factor in Russia's energy production as early as expected by some. And it is still unjustified to state without reservation that the Russian Arctic continental shelf is set to become a major arena for the international oil industry in the course of the next ten years.

In addition to the costs and uncertainties linked to implementing the currently presented oil projects the total resource base does not look as attractive as it did only a few years ago. Some 80 per cent is expected to be natural gas (Gazprom, 2012). Arctic offshore was expected to become a major source of gas supplies, but with the drastic changes in natural gas markets sparked by emergence of several new LNG projects and then the rise of unconventional gas, this no longer seems inevitable. Gas prices are not expected to grow as before, and the profitability of Arctic offshore gas projects has been questioned. That has implications for all offshore exploration, since discoveries of gas (even if the primary goal is to find oil) are now less likely to lead to industrial development. In turn, the commercial risks will be higher.

---

8   Rosneft press release 13 February 2013. www.rosneft.ru.

East Siberia

Besides the Arctic, East Siberia has been identified as a future energy resource base of Russia. For many years now, scattered exploration and geological theories have indicated a vast resource potential, but actual development has been very limited, for understandable reasons. Long distances from existing infrastructure and extreme climatic conditions characterize many potential oil-producing districts there. Several projects are under various stages of development (see Table 2.3). The cost factor is a serious concern, with some estimates indicating that cost per unit of oil production capacity will be on average 2.5 times higher than for new fields in West Siberia. Even an optimistic assessment of the output potential does not put it above 45 mill. tons a year by 2020–2030, as compared to 19.5 mill tons in 2010 (Prischepa and Podol'ski, 2012). Interestingly, the capacity of the East Siberia–Pacific oil pipeline was first planned to reach 80 million tons per year (Government of Russian Federation, 2009), but present concrete plans apparently do not go beyond 50 million tons.[9] A considerable part of that oil will have to come from West Siberia. Thus, there is already a tendency to play down the potential of East Siberia, and the most promising parts, including the giant Vankor field, are actually geologically an extension of West Siberia.

Industry Structure

The realization that East Siberia will be very costly to develop, and the accompanying huge investment requirements, have become a major issue in discussions about the structure of the oil industry in Russia. In any country, there are risks entailed in developing the mineral sector, and it seems clear that in Russia the risks are increasing. At the same time, there are elements in the institutional framework, both taxation and the licensing system, that are dysfunctional when it comes to risk-taking.

But in addition to inadequate formal framework conditions, there is a further question: do Russian oil companies have features, or characteristics, that make them inclined not to take risks? On the general level, we will argue that Russia's petroleum companies – state-controlled as well as private – live under 'soft institutional constraints'. This is a wider term than 'soft budget constraints' coined by Kornai (1980), denoting the ability of enterprises in the centrally planned economy to exceed budgetary limits and be compensated for over-spending. Soft institutional constraints imply that enterprises are able to manipulate their framework conditions to increase their profit, whether through tax concessions or legal provisions or specific conditions for new development projects. The emergence of soft institutional constraints must be seen in connection with the political and economic system that has developed in Russia. The system is oriented towards

9   http://www.transneft.ru/projects/119/10025/.

Table 2.3 Oil projects in East Siberia

Project	Region	Company	Implementation	Annual production capacity (mill. tons)	Estimated total investments (USD)
Vankor	Krasnoyarsk kray	Rosneft (Vankorneft)	2003–37	25.5	26
Suzunskoe	Krasnoyarsk kray	TNK-BP (Suzun)	2011–20	1.7–2	2
Tagulskoe	Krasnoyarsk kray	TNK-BP (Tagulskoe)	2011–20	4.5–5.6	3
Yurubcheno Takhomskoe – 1st phase	Krasnoyarsk kray	Rosneft (VSNK)	2011–14	10	3
Kuombinskoe and Tersko-Komovskoe	Krasnoyarsk kray	Slavneft (Slavneft-Krasnoyarsk-neftegaz)	2010–39	10–11	8
Verkhnechonskoe	Irkutsk obl.	TNK-BP/Rosneft (Vekhnechonsk-neftegaz)	2009–20	7	5
Yaraktinskoe	Irkutsk obl.	Irkutskaya neftyanaya kompaniya	2005–33	2.4	2
Talakanskoe	Sakha rep. (Yakutiya)	Surgutneftegaz	2009–40	7	13
Sredne-botuobinskoe	Sakha rep. (Yakutiya)	Rosneft (Taas-Yuryakhneftegazo-dobycha)	2014–40	4.5–6	3

Source: Kryukov (2012) based on scattered company information.

support for big financial and industrial structures. Their leaders – who are fairly limited in number – interact with the authorities on a personal level, and the authorities participate directly or indirectly in many companies. Companies are to a large extent controlled by individuals, as opposed to publicly traded companies with many owners. In terms of economics, the system is oriented toward rapid pay-offs from investments and existing assets. We believe this combination of political and economic characteristics logically leads to priority for large-scale projects with 'guaranteed' returns and minimal economic risks.

In most countries with a maturing petroleum sector, a diversified industry structure is usually regarded as a precondition for effective resource management. Small, specialized companies take care of tail production from fields no longer of interest to the big companies, while specialized exploration companies can venture into new areas with particular challenges, turning discoveries over to regular production companies.

In Russia, however, despite the changes in the resource base, the petroleum sector has remained totally dominated by big vertically integrated companies – indeed, this is in line with official policy. The five leading oil companies stand for 85 per cent of Russia's output. In all there are eight vertically integrated oil companies. There are also about 150 other, relatively small, companies – these include companies with Russian as well as foreign owners, and mixed types. But despite their numbers, they do not play an important role in Russian energy policy. The developments evident in the resource base call for pluralism in approaches and solutions to exploration as well as production. The big companies can only offer a limited choice. Also the considerable unconventional oil resources might be easier to access with a more diversified industry structure, due to the high risks as well as the need for specialized technological skills. According to Rosnedra, the subsoil agency, the most promising formation in West Siberia may hold between 25 and 50 bill. tons of recoverable resources (*Platts*, 12 October 2012). In the USA, such resources have typically been developed by relatively small, independent companies. In Russia, a further problem comes from the limited competition between companies for licenses in many regions. Altogether this means that the costs of developing the resource base are much higher than they could have been with a more varied and competitive industry structure.

The most striking development in recent years has nevertheless been the de-privatization of the Russian oil industry. In the 1990s, several privately-owned companies were built up on the basis of structures inherited from the Soviet state industry. By 2004, the five largest companies were all private. State-owned Rosneft belonged in the group of second-tier companies, providing less than five per cent of Russian production. But from the same year, the trend was reversed. State-dominated companies took over some of the private ones – notable here were Gazprom's takeover of Sibneft and Rosneft's *de facto* takeover of Yukos. In parallel, the state-owned companies got preferential treatment in licensing processes. Thus the balance between private and state-controlled companies had already changed considerably by the time Rosneft bought out the foreign as well as

Russian shareholders in TNK-BP in 2012 as depicted in Table 2.4. With inclusion of the latter company's production, Rosneft will represent some 37 per cent of Russia's output. Gazprom's oil subsidiary Gazprom Neft takes care of another six per cent.

That a state-owned or state-dominated oil company in an oil producing country has a dominant position is nothing unique to Russia. In Norway, for example, Statoil is operator for about 72 per cent of the country's oil and gas production, whereas the company's equity and entitlement production in 2011 constituted approximately 35 per cent of total Norwegian output (Statoil, 2012; NPD, 2012). Thus it might be argued that Rosneft's new dominating position is in line with the situation in

Table 2.4   Composition of Russian oil producers 2004 and 2011 (mill. tons)

	2004	2011
Lukoil	84.07	85.32
Yukos	85.68	
Surgutneftegaz	59.62	60.78
TNK-(BP)	49.74	72.63
Sidanko	20.77	
Rosneft	21.6	114.49
Sibneft/Gazprom Neft	33.99	30.29
Slavneft	22.01	18.08
Tatneft	25.09	26.19
Bashneft	12.07	15.10
Russneft	8.79	13.63
Gazprom	11.96	14.52
Others	25.8	59.91

Source: *Neft' i Kapital*, 1–2, 2005 and 1–2, 2012.

other petroleum-producing countries. But this argument evades the question of what sort of structure Rosneft really is and is becoming. The Russian petroleum sector is characterized by soft institutional constraints and constant bilateral negotiations between companies and authorities over basic rules of the game. The state-dominated companies have obvious advantages in this situation, with very short distances between authorities and companies and sometimes personnel overlap. The most glaring example is Igor Sechin's dual role as deputy prime minister in charge of the fuel and energy complex, and at the same time chairman of the board of Rosneft. This constellation was brought to an end by President Medvedev in 2011, but a new combination emerged after Putin's return to the presidency. Now Sechin became president of Rosneft and at the same time secretary of the president's new commission for strategic development of the fuel and energy complex, ostensibly coordinating the activities of various companies in the field.[10]

For Putin it seems that the rationale behind Rosneft's expansion is to concentrate resources in a company responsive to national strategies that is big enough to handle the investment challenges in the sector. Privately-owned companies do not take a sufficiently long-term view. But is the policy of favouring state-controlled companies likely to achieve these goals? Rosneft now comes across as a more modern company than, for example, Gazprom. It has foreign part-owners – and will get more as BP's stake is increased to 20 per cent, and a portion of its shares are traded at stock exchanges. Rosneft has a board of directors with several

---

10   http://kremlin.ru/news/15656.

independent members, including non-Russians, and it has also for some years recruited experienced executives from abroad.

Nevertheless, we hold that the peculiar relationship between state and company in Russia remains, and that soft institutional constraints and weak market structures characterize the environment surrounding the oil and gas sector. This makes a dominant position for one company more risky than it would be in a country with stable and clear framework conditions and a functioning market for services and equipment.

Concluding Remarks

We have noted several alarming tendencies in the Russian oil resource base: decreasing size of discoveries, falling recovery rate, uncertainties about the volumes and qualities of reserves ready to be developed. These trends – which have been noticeable over several years – are now widely recognized also by top Russian officials. The most immediate concern for the government is, however, to secure state revenues. This is becoming increasingly challenging, as the price of oil has fallen in real terms compared with a few years ago, while production costs continue to rise.

It is increasingly recognized that serious changes in the framework conditions are required, particularly in taxation, in order to stem the negative developments in the oil fields under production as well as to explore and develop resources in new regions that can secure longer-term output. But, as we have shown here, making Russia's taxation system more conducive to rational development of resources is easier said than done, and tax reform involves the risk of a shortfall in government revenues, at least over the short term.

The industry structure as it has evolved since the break-up of the Soviet Union, with concentration of activities and assets in a limited number of companies, answers to a perceived state need for control, but is in increasing conflict with developments in the resource base, as well as the possibility of exploiting significant unconventional resources. Whereas the tax reform issue is now being openly debated, there has been much less public discussion about the structure of the industry. One reason is that the present structure is linked to the personal ambitions and economic interests of powerful actors in society; another explanation is that almost all the proven reserves have been licensed to the industry already. This means that the scope for a new, balanced resource management policy is limited. Such a policy would typically entail combining licenses for 'easy' fields with more complicated projects, as a way of stimulating interest in the latter. Nevertheless, the authorities *can* do something to stimulate new companies in specific regions, and we believe they will, in the same way as the framework conditions for foreign participation on the continental shelf have been improved.

Of course, there are counter-arguments and resistance to radical change in policies and industry structure. But our conclusion is that the developments in the resource base are so severe that some change will have to come. However, this may well come too late to avert a fall in oil production in Russia.

References

Agranovich, M. 2008. Innovatsii pozvolyat effektivno ispol'zovat' nashi resursy [Innovations permit effective use of our resources]. Interview with Sergei Mazurenko. *Rossiyskaya Gazeta*. 12 November. Available at: http://www. rg.ru/2008/11/12/othodi.html [accessed 15 November 2012].

Barsukov, Y., V. Visloguzov and D. Butrin 2012. Shel'f razbirayut na chastnoe [Shelf is disassembling]. *Kommersant.ru*, 3 August.

Bokserman, A. A. 2011. Rossii nuzhna effektivnaya strategiya razvitiya neftyanoj otrasli [Russia needs an efficient oil industry development strategy], *Burenie i neft*, no. 2. Available at: http://burneft.ru/archive/issues/2011-02/7 [accessed 15 November 2012].

BP Statistical Review of World Energy 2012. London, UK: BP p.l.c. Available at: bp.com/statisticalreview.

Bushuev, V. V. and A. I. Gromov (eds). 2007. *Toplivno-energeticheskiy kompleks Rossii 2000–2006*. Moscow: Energiya.

Dienes, L. 2004. Observations on the problematic potential of Russian oil and the complexities of Siberia, *Eurasian Geography and Economics*, 45(5), 319–45.

Donskoy, S. 2010. Speech at ONS in Stavanger 26 August 2010, (Russian) Ministry of Natural Resources and Ecology. Available at: www.mnr.gov.ru/ part/?act=more&id=7119&pid=11 [accessed 15 November 2012].

Gazprom 2012. *Russian continental shelf*. Available at: http://www.gazprom.com/ about/production/projects/deposits/shelf/ [accessed 15 November 2012].

Government of the Russian Federation 2009. *Energeticheskaya strategiya Rossii na period do 2030 goda [Energy Strategy of Russia through 2030]* (Decree of the Government of RF # 1715-p, 13 November 2009), Moscow: Government of the Russian Federation.

Khalimov, K. E. 2003. *Evolyutsiya otechestvennoy klassifikatsii zapasov nefti i gaza* [Evolution of the domestic classification of oil and gas reserves]. Moscow: Nedra.

Khramov, D. 2012. *Deputy Natural Resources Minister D. Khramov in interview with 'Golos Rossii'*. Available at: http://www.mnr.gov.ru/press-service/publications/ detail.php?ID=128451&sphrase_id=128623 [accessed 15 November 2012].

Kornai, J. 1980. *Economics of Shortage*. Amsterdam: North-Holland.

Kryukov, V. 2012. Institutsional'nye baryery razvitiya neftegazovogo sektora Rossii (na primere Vostochnoy Sibiri) [Institutional barriers for the development of the Russian petroleum sector (exemplified by East Siberia)], *Zhurnal novoy ekonomicheskoy assotsiatsii (NEA)*, 4(16), 151–7.

Kryukov, V. and A. Moe. 2007. Russia's oil industry: risk aversion in a risk-prone environment. *Eurasian Geography and Economics*, 48(3), 341–57.

Maksimov, V. M. 2011. O sovremennom sostoyanii neftedobychi, koeffitsiente izvlecheniya nefti i metodakh uvelicheniya nefteotdachi [About the current state of oil production, rate of recovery and methods to increase oil recovery], *Burenie i neft*. Available at: http://burneft.ru/archive/issues/2011-02/6 [accessed 15 November 2012].

Mazneva, Y. 2011. L'gota Millera [Miller's privilege], *Vedomosti*, 8 June. Available at: http://www.vedomosti.ru/newspaper/article/261701/lgota_millera [accessed 15 November 2012].

Melnikov, K. 2012. Gosudarstvo ne dobralos' do zapasov. Vvedenie ikh novoj klassifikatsii otkladyvaetsya [The government did not reach the reserves. Introduction of their new classification is set aside], *Kommersant*, 30 January. Available at: http://www.kommersant.ru/doc/1861759 [accessed 15 November 2012].

Moe, A. 2010. Russian and Norwegian petroleum strategies in the Barents Sea. *Arctic Review on Law and Politics*, 1(2), 225–48.

Moe, A. and E. Wilson Rowe 2009. Northern offshore oil and gas resources: policy challenges and approaches, in *Russia and the North*, edited by E. Wilson Rowe. Ottawa: University of Ottawa Press, 107–28.

MPR (Russian Ministry of Natural Resources and Ecology) 2011. *O sostoyanii i ispol'zovanii mineral'no-syrevykh resursov RF v 2010 godu. Gosudarstvenny doklad [About state and usage of mineral and raw material resources in RF in 2010. State report]*. Available at: http://www.mnr.gov.ru/regulatory/detail.php?ID=128810 [accessed 15 November 2012].

Muslimov, R. 2009. KIN – ego proshloe, nastoyaschee i buduschee na mestorozhdeniyakh Rossii [The oil recovery rate – its past, present and future in Russian fields], *Burenie i neft'*, 2.

Muslimov, R. 2012. *Nefteotdacha: proshloe, nastoyaschee, buduschee* [Oil recovery: past, present, future], Kazan: FEN publishers, Academy of Sciences of the Republic of Tatarstan.

Muslimov, R. 2012. Chto meshaet vnedryat' innovatsii v neftyanke? [What is in the way of innovations in petroleum sector?], *Neftegaz.ru*, 8, 23–9.

Novak A. 2012. Novak: RF poteryaet chetvert' nefti pri suschestvuyschikh neftyanikh nalogov [Novak: RF will lose a quarter of its oil with existing taxes], *Russian Business Consulting*]. Available at: http://top.rbc.ru/economics/21/06/2012/656101.shtml [accessed 15 November 2012].

NPD (Norwegian Petroleum Directorate) and Ministry of Petroleum and Energy 2012. *Facts 2012 – The Norwegian Petroleum Sector*. Available at: http://www.npd.no/en/Publications/Facts/Facts-2012/ [accessed 15 November 2012].

Orlov, V. P. 2008. Realii i problemy otechestvennoy geologorazvedki [Realities and problems of national geological exploration], *Mineral'nye resursy Rossii—ekonomika i upravleniya*, 3, 2–6.

Overland, I., J. Godzimirski, L. P. Lunden and D. Fjaertoft. 2012. Rosneft's offshore partnerships: the re-opening of the Russian petroleum frontier? *Polar Record*, 49(2), 140–53. Available at: http://journals.cambridge.org/abstract_S0032247412000137 [accessed 15 November 2012].

Pavlenko, V. 2011. Call of the Arctic. *Oil of Russia*, 3, 46.

Popov, A. 2012. *Speech of the head of Rosnedra to the VII All-Russian Congress of Geologists*. Available at: http://www.rosnedra.com/article/6158.html [accessed 15 November 2012].

Prischepa, O. and Y. Podol'skiy. 2012. 420 mln ton neft k 2030 godu? [420 mln tons of oil by 2030?], *Neftegazovaya vertikal'*, 8, 44–53.

Sadovnik, P. V. 2006. Osnovnye itogi raboty Rosnedra v 2005 i zadachi na 2006 g. v chasti uglevodorodnogo syr'ya i podzemnykh vod [Main results of the work of Rosnedra in 2005 and tasks for 2006 in the area of hydrocarbon resources and sub-soil water]. *Mineral'nye resursy Rossii—ekonomika i upravleniya*, 3, 13–21.

Shmatko, S. I. 2010. Doklad po voprosu General'noy skhemy razvitiya neftyanoy otrasli na period do 2020 goda [Report on the question of the general scheme for development of the oil sector until 2020. 28 October 2010. Available at: http://minenergo.gov.ru/press/doklady/?PAGEN_1=2.

Statoil 2012. *Annual report 2011*. Available at: http://www.statoil.com/AnnualReport2011/en/Pages/frontpage.aspx.

Vygon, G. V. 2010. *Okhrana okruzhayushchey sredy i povyshene neftootdachi [Environmental protection and rise in oil recovery]*. Available at: http://www.imemo.ru/ru/conf/2010/161110/Vigon.pdf [accessed 15 November 2012].

Chapter 3

After the Crisis:
New Market Conditions?

Tatiana Mitrova

Introduction

Global energy markets have changed radically during the recent economic crisis. High oil prices, weak economic performance and progress in shale oil production have completely reshaped the global oil market, slowing down oil demand in the OECD countries: after peaking in 2005–2007, European, North American and OECD Asia oil demand has not yet regained pre-crisis levels.

Similarly, the shale gas revolution, progress in LNG technologies, the economic recession and several other factors have completely closed the North American gas market for imports and reduced dramatically European gas import requirements, and in the future might lead to lower gas import needs in the Asia-Pacific markets as well.

All these changes have taken place just as Russia, which had been developing its energy sector based on a strategy of extensive production and expanding export volumes, has entered into an era of new and far more expensive upstream and transportation projects. In the new environment with much lower growth rates in demand and far stronger competition (a 'buyers' market'), Russia for the first time has to compete without the advantage of cheap energy resources.

This chapter investigates the situation immediately after the 2008/2009 crisis and potential future development of the chief branches of the Russian energy sector. Main challenges in the new market conditions for the Russian oil, gas and coal industries are examined, as well as the changes in the Russian hydrocarbon export policy towards Europe and Asia intended to help Russia adapt to the new, less favourable external market environment.

The Crisis and its Impact on Russian Energy Sector

After a steep decline during the 1990s, the Russian economy and the energy sector had recovered to nearly pre-reform levels by 2008. However, the 2008 world crisis caused new losses which will take longer to repair.

Even before the global economic downturn, the Russian hydrocarbon sector found itself at a crossroads. For years, Russia had been able to reap the benefits of Soviet-era investments in oil and gas, but those resources have begun to decline, making new large-scale investment necessary (Zagashvili, 2010). The global crisis has greatly complicated implementation of such plans, and policy responses have not kept up with the reforms needed to ensure sustainable growth in the energy industry (Mitrova, 2009).

Economic recovery after the difficult transitional period in the 1990s and later the rising prices for hydrocarbons provided a huge stimulus to growth in production and export, and pushed forward many ambitious plans. Russia's oil and gas industry, it seemed, had successfully managed the phase of post-Soviet institutional reshaping (Astrov, 2010; Sergeyev, 2010b). Between 1998 and 2008, Russia showed the greatest incremental growth in oil and gas production in the world (an additional 187 million tons of oil and 65 bcm of gas in 1998–2008). Oil export more than doubled, with Russian oil production reaching the level of Saudi Arabia. For a while this gave a boost to the concept of Russia as an 'energy superpower', as the huge revenues generated by sales of oil and gas bolstered Russia's self-confidence and belief in a brighter future.

But then, starting in 2005, the increase in Russian crude oil production began to slow down; in 2008 it even fell by 0.6 per cent. There were several reasons for this. Firstly, production efficiency was reduced due to the declining quality of the industry resource base and difficulties in developing new oil fields. The oil companies lacked incentives for stepping up production because of the high tax burden as well as the sharp decline in world oil prices in September–December 2008, which led to upstream losses in the final quarter of that year. By February 2009, oil production in Russia had fallen by 5.7 per cent as compared to one year earlier. Domestic demand for oil products dropped by more than 13 per cent. With domestic demand simply no longer there, the oil companies sought to export as much as they could. Moreover, domestic product prices failed to recover toward export netbacks after the 2009 devaluation of the rouble, which made exports more attractive. Thus, by 2009 production was decreasing in Russia, whereas exports were on the rise. In 2008 and 2009 the government provided considerable decreases in the tax burden. Oil company profits have risen from their low point, thanks to lower export duties and rouble devaluation. But the taxation changes introduced in 2009 were still insufficient to enable complete recovery (Sergeyev, 2010a).

Similarly, since the last quarter of 2008, the gas sector in Russia has experienced a downturn in terms of prices, demand and production, with production dropping to 2000 levels and export – to 2004 levels. In 2009, gas production fell to nearly 2000 levels, totalling 583.6 bcm as against 664 bcm in 2008 (−12.1 per cent). Several factors were responsible: reduction in domestic demand (from the energy sector, machine industry and other sectors of the economy) and, even more important, reduced demand from the external market; high gas export prices in the first six months of the year; and a halt in the transit of Russian gas through Ukraine in January 2009 (Stern, 2009; Mitrova, 2011a). Gazprom was hardest hit, its gas production dropping in the first quarters of 2009 by 18 per cent year-on-year. The global economic downturn lowered export demand for gas – in 2009 gas exports were down by an extraordinary 12 per cent. The drop in export prices meant a decline in revenues, and lower incentives for future investments and energy saving (Mitrova, 2011a).

Also the rest of Russia's energy industry was heavily affected by the crisis. Investments in new projects slowed down, the interaction between the electricity generation sector and the rest of the economy was complicated by a substantial drop in industrial production. Coal production suffered particularly, with some companies reporting an almost 60 per cent reduction in output year-on-year in January/February 2009 (Tarazanov, 2009a).

The crisis served to worsen the prospects of the Russian energy sector, bringing stagnation and adversely affecting its further development. The crisis hit the country in the midst of several major energy reforms, and those not been completed by the final quarter of 2008 were endangered by the shifts then underway in the energy market.

In particular, the economic crisis can be said to have had three clear effects on the Russian economy as a whole: loss of external funding, lower demand for Russian products and services, and a fall in export revenues. All three are closely linked to the energy sector. Firstly, the loss of external funding was one of the earliest recognizable features of the world economic crisis, as investor confidence fell and many actors faced severe liquidity problems. This was deeply felt by Russia's electricity sector in particular, which had been embarking on costly reforms aimed at increasing the country's generation capacity. These reforms were intended to help Russia to cope with future rises in demand, to lessen its overdependence on gas power and to free up further capacity for export. The crisis made it harder to attract external funding for this project, which have struggled to fulfil its targets. The lack of foreign loans and the necessity of meeting commitments for loans on securities under toughening stipulations (caused by the slump in securities prices) has affected loan terms, intensifying the negative tendencies in investment dynamic in all productive sectors in Russia (Mitrova, 2011b).

The problem of lower demand has been felt throughout the Russian energy industry, which has reacted by cutting production. In addition it has forced producers to adapt to new market conditions. Oil and coal producers have lobbied for lower taxes, and coal producers have searched for new markets (Grigoryev, 2009).

Lower demand has meant lower prices, with producers seeking to retain their market shares by appealing to the limited number of buyers. Some of the greatest effects have been seen in the coal industry. Russian coal became less competitive on the global market when, in 2008, the price on international markets dropped below the combined costs of production, transport and taxation from Russian mines (Tarazanov, 2009b).

Another huge problem created by the crisis has been that in this extremely unstable situation the government has been less active in promoting energy-sector reforms and price deregulation for the power and gas sector. Considering the rates of inflation, the prices in the regulated sectors were actually frozen in order to support competitiveness in industrial production and avoid social protests from household consumers. This has meant fewer incentives for implementing structural changes in Russia's energy sector. This in turn may lead to problems in meeting energy efficiency goals (Smirnov, Yanchenko and Reshetilo, 2011).

Generally speaking, the crisis has worsened the prospects of the Russian energy sector, bringing stagnation and adversely affecting the stimulus for its later development. Slower growth in domestic fuel prices could decrease the pressure on domestic industrial producers, as well as obstructing energy-saving initiatives and delaying changes in the structure of the economy and the energy sector. At the same time, lower demand at the traditional export markets and growing competition with the other suppliers are creating external limits for further Russian energy export growth. This, coupled with the decreasing efficiency of new field development, may mean that Russian energy exports will slow down in the longer term (Mitrova, 2011b).

Although Russia has expected the crisis to end quickly, any longer decline of demand for energy could result in more problems for the struggling economy. This is quite understandable in the case of a country with such a dominant energy sector: energy is responsible for creating 30 per cent of Russia's GDP and represents a huge share of the government's revenues (Figure 3.1). In the period 2000 to 2008 it sometimes seemed that the energy sector was supporting and pushing forward the other parts of the economy, through subsidized low-price gas and multiplication mechanisms created by investments in the energy sector (Tabata, 2009). Then the 2008/2009 economic crisis revealed the threats and risks of this heavy overdependence on hydrocarbons.

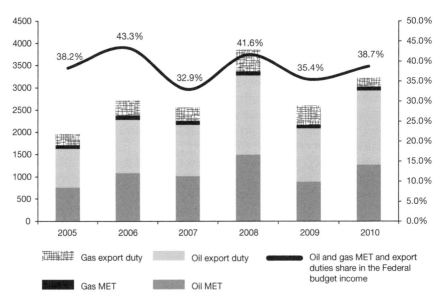

Figure 3.1   Budget incomes from oil and gas, bill. rouble and shares
Source: Federal Treasury.

The only solution for Russia is to adapt to the changed circumstances. The government needs to adopt new fiscal policies in order to ease the fiscal burden, create more attractive rules of the game for the investors as well as adjusting to the post-crisis world by enacting reforms to make Russian energy sector more competitive in the future. So far these long-awaited changes have not materialized.

Russia has responded to the crisis by trying to adhere to its original plans, apparently betting on the crisis being brief enough to allow the country to re-emerge strengthened afterwards. These approaches are formulated in the official *Energy Strategy of the Russian Federation until 2030* (Government of Russia, 2009), or ES-2030, which was adopted during the most difficult phase of the crisis in November 2009. This document focuses primarily on the domestic agenda – modernization of the energy and non-energy sectors; and reducing dependence on energy exports, energy intensity, and the share of gas in power generation – but the main priorities of international cooperation in the field of energy are also discussed.

Energy Strategy 2030

Russia has clearly recognized its overdependence on energy exports, and the 2008/2009 crisis has again highlighted this weakness in the country's economy. Formally, the major target of the Energy Strategy (ES-2030) is to reduce the share of the fuel and energy complex in Russia's GDP and the share of energy resources in exports (by at least 1.7 times). The share of capital investments in the fuel and energy complex (in percentage of GDP) is meant to decline at least 1.4-fold and the share in overall investments be halved, as is GDP energy intensity – these are the major domestic targets of the ES-2030. In addition, the share of energy resources in export is to fall from 68 per cent in 2008 to 57 per cent by 2030. It is also planned that more of primary energy resources will be consumed within Russia – 61.3 per cent in 2030, compared with 53.2 per cent in 2008. It is also expected that Russian energy mix will undergo some changes: the share of gas in the mix is to decrease from 53 per cent in 2008 to 45 per cent by 2030, the share of oil and other resources is to increase from 19 per cent to 22 and from 10 per cent to 14 per cent respectively, while the share of coal is to remain at almost the same level – 18 per cent in 2008 and 19 per cent in 2030.

The Energy Strategy plans for a step-by-step development of the energy sector in the post-crisis period. The document envisions two scenarios – basic and innovative. Strategy implementation is divided into three stages, to be completed in 2015, 2022 and 2030. In the first stage (2013 to 2015) it is planned to overcome the crisis in the energy sector and create conditions for growth. In the second stage (2015 to 2022), a general increase in energy efficiency is expected through innovative development in the energy sector. In the final stage, 2022 to 2030, the focus will be on energy efficiency and on creating conditions for the development of non-fossil energy sources (see Strukova, 2011). The Energy Strategy also aims at reducing the share of gas in total energy consumption in the Russian Federation from 52 per cent to 47 per cent, and increasing non-fossil fuel energy from the current 10 per cent to 14 per cent, giving a more balanced energy mix.

As to external targets, it should be noted that Russia, with only some 3 per cent of world population and GDP, holds 6 per cent of global oil reserves, 24 per cent of global gas reserves and produces roughly 12 per cent of all global primary energy. The country exports almost half of the oil, one third of the coal and one third of all the gas that it produces. We should also recall the 'hidden energy export' in the form of energy-intensive products like aluminium and fertilizers, making Russia an extremely important player on the global energy scene (Sergeyev, 2010b). On the other hand, it must be stressed that the idea of Russia as an 'energy superpower' no longer figures on the national political agenda. The Russian authorities have recognized that this notion is useless and may even be harmful to the country's image (Romanova, 2010). But even without these ambitions, Russia's foreign energy strategy is highly relevant for all other participants on the market (see also Shadrina, 2010). Let us have a look at the main external priorities of the ES-2030.

First comes the need for Russia to *focus on its national interests in the international energy regulation system*. In Russia the processes of internationalizing energy legislation have been slow, but are inevitable. What characterizes Russia are the deep differences in the economic development of the various parts of its energy sector.

The second strategic priority is *diversification of the product structure of exports*. The share of exported crude oil should decline, whereas the share of oil products in total Russian export should increase significantly. Also the share of LNG in gas export should grow, reaching 14–15 per cent by 2030. For historical and geographical reasons Russia did not enter the LNG market until 2005, and large-scale LNG production began only in February 2009 with the Sakhalin-2 project. The main advantage of LNG is that deliveries can reach consumers in wide-ranging parts of the world. For Russia, this means being able to reach out to the Atlantic basin (from the Arctic projects) as well as the Pacific basin (from Sakhalin and Vladivostok-LNG). Moreover, LNG offers potential access to new markets and improved security of supply to existing markets – a concern which remains central to the nation's external energy strategy (Mitrova, 2011c).

The third priority is the *diversification of export markets*, with the main focus on the Asian market. Russia's hydrocarbons industry today is oriented firmly towards the West. In 2008, only 14 million tons of crude oil were exported to the Asia-Pacific region (APR), against a total export volume to all destinations of 243 million tons – with a massive 185 million tons going to Europe. The picture is even clearer in gas, with no major exports to Asia until 2009.

However, there is very limited potential for further export growth in the Western direction: European oil and gas markets are clearly in stagnation. The situation is aggravated by the fact that EU–Russia energy relationships are deteriorating: they have become so complicated, highly charged and politicized that actions often go against economic logic. The main challenge in EU–Russia energy cooperation remains the lack of understanding on the central issue of energy security. Europe wants Russia to increase investments and radically increase the participation of Western companies in its energy production. The Russian response has been based on the concept of strategic reserves, with restrictions on foreign participation in production and Gazprom's monopoly on transport and export (Romanova, 2010; Sherr, 2010). The disputes in 2011 and 2012 are due first of all to the anti-monopoly investigations against Gazprom and pressure on Russian long-term oil indexed contracts, showing how two different logics are at play. Russia, a main producer that needs to invest in development of new more expensive fields, is interested in long-term, predictable contracts that give it long-term security when making investment decisions; European gas buyers, following the logic of a consumer who is in addition hit by a deep economic crisis, want to cut the immediate costs in a situation when they may hope for possible supplies of cheaper gas coming from other directions, be it the USA or the Middle East. All these conflicts, combined with new global market conditions, could make Russia more reluctant to invest in the new expensive supply projects to Europe – the Shtokman abandonment in August 2012 being one of the first signs of this new strategy. However, Russia does not foresee Europe ceasing to be the main destination for its hydrocarbon exports – only that the investment focus is expected to move from incremental supplies to the West from the European part of Russia and Western Siberian fields into Eastern Siberia and Far East development, which has also very strong domestic logics and socioeconomic implications.

Russia now expects Asia to play a more important role (Averre, 2010; Aalto, 2007). Especially China's continued economic growth even during the period of severe global recession seems set to boost Russian exports. The ES-2030 predicts a gradual increase in oil exports to Asia of 70–80 million tons by 2030, and gas exports are expected to rise to 70–75 bcm, reaching approximately 20 per cent of Russia's total gas export. The challenge is to raise the share of energy export to Asian markets to 26–27 per cent of total energy export, accompanied by sustained volumes of exports to European markets.

According to the ES-2030, in the future total Russian exports should increase by 15–20 per cent, and gas exports will grow even more. Exports to the APR are expected to increase up to six-fold, whereas exports to Europe and the CIS are set to decline (Makarov, Grigoriev, Mitrova, 2012: 152).

This is no easy task. In order to develop its exports to the East, Russia must first resolve various domestic issues: it will have to develop several complicated greenfield projects and build huge new infrastructure in remote areas. Enormous investments have to be attracted, and in turn necessitating a completely different regulatory environment. There are also some questions related to the external markets: both China and OECD Asian countries are very active in securing their future supplies, assuming that various new suppliers will enter this premium market (like the USA, Canada, Australia – countries where new oil and gas exports are provided mainly by massive expansion in unconventional production). Now the challenge facing Russia is whether it can be an energy giant and

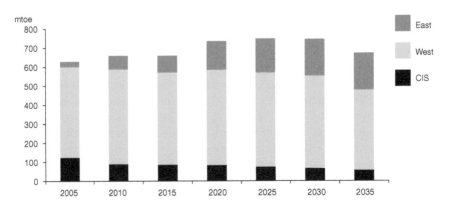

Figure 3.2   Energy exports by direction 2005–2035 in mtoe

Source: Global and Russian Energy Outlook until 2035. ERI RAS-REA. Moscow, 2012.

successfully compete on two continents simultaneously: and that means Russia will need not only its natural resources, but also the requisite infrastructure and policy framework (Fernandez, 2009a; 2009b).

Since 2006 disputes between Russia and some of its neighbours have caused disruptions in fuel supplies to the EU. Consequently another element of the Russian Energy Strategy is *the diversification of export routes* in order to increase security of supply and reduce transit risks. Several huge projects are relevant here – the Baltic Pipeline System-2, Burgas–Alexandroupolis, Samsun–Ceyhan for oil transport; and Nord Stream and South Stream for gas. These pipeline projects are in varying phases of realization. BPC-2 is under construction, whereas two competing projects that would bypass the Strait of Bosporus – Burgas–Alexandroupolis and Samsun–Ceyhan – are in the phase of preliminary negotiations with little hope of success. Work on the first of the two Nord Stream pipelines was completed in 2011 after Denmark, Sweden, Finland and Germany had granted permission and 29 banks confirmed their readiness to participate in financing. The second line was commissioned in October 2012. South Stream is at a less advanced stage; nevertheless some success was achieved in 2008/2009, when intergovernmental cooperation agreements were signed with Bulgaria, Hungary, Greece and Serbia, and new long-term contracts for gas supplies were signed with key gas customers.

A further priority in Russia's external energy strategy is providing stable conditions on the export markets, including security of demand and reasonable prices for Russian energy resources (Mitrova, 2011c). Particular attention is paid to long-term take-or-pay contracts, as they are expected to ensure high volumes of sales and stability in gas market relations. Russia has demonstrated quite a rigid approach towards all attempts to review the traditional gas export system; in 2012, all responsibilities for these negotiations were officially transferred from the corporate to the state level.

At the same time, Russian energy policy *seeks to strengthen the position of Russian energy companies abroad*. The aim is for one Russian energy company to rank among the world's three leading energy companies, and two of them among the top ten.

Another major priority is to provide efficient international cooperation on risky and challenging projects in Russia, including offshore projects in the Arctic (Mitrova, 2011d). The share of foreign direct investments in total investments in the Russian energy sector should be at least 12 per cent, while Russian companies should also participate actively in projects abroad. While such cooperation in the sphere of oil exploration and production seems quite successful (Rosneft's deals with BP, ExxonMobil, Statoil and ENI are probably the most well-known and promising projects), the gas industry has been less fortunate, as Shtokman has been postponed for an undetermined period and Yamal-LNG with Total seems to be the only large-scale Arctic project with an international partner.

Oil Industry

Currently the Russian region with the highest production of oil is the Tyumen region in West Siberia, more specifically in the Khanty-Mansy Autonomous Okrug (KhMAO). In 2030 it will remain the main region of production of crude oil. Eastern Siberia has not yet played any major role in Russian oil production. Companies working in this difficult region have encountered a range of problems, such as high geological risks and delays in infrastructure development, high capital and operational costs, the challenge of filling the East Siberia–Pacific Ocean (ESPO) pipeline, while the government's plans on petroleum licensing in the region have lagged far behind. But these Eastern regions are expected to become increasingly important over the next 20 years, eventually producing around one fifth of Russia's oil and gas. Production is also expected to grow in the country's Far East, North and Northwest and Caspian regions (Makarov, Grigoriev and Mitrova, 2012: 153). By 2030 Eastern Siberia is expected to produce 65–69 million tons, Northwestern Russia 42–43 million tons, the Volga region 34–36 million tons, the Urals 25–29 million tons, the Caucasus and the Caspian region 21–22 million tons and the Far East 32–33 million tons. The Tyumen region is set to remain the leading producer, reaching some 291 to 292 million tons.

As hydrocarbons exports to the Far East are beginning from a relatively modest level, one vital question for policy makers is the matter of new export routes in the Russian East. The state oil transport company Transneft has already constructed the first phase of the ESPO pipeline, the one reaching Chinese border, and is finalizing the construction of the second phase, to bring oil to the Pacific coast. As a result, the share of exports to Europe by 2030 is foreseen to decrease from 80 to 65 per cent, with the share of exports to the Asia- Pacific region rising accordingly (Fernandez, 2009b).

Here we should note one radical change: once the crisis is over, Russia will not be able to increase substantially its production and export much more oil.

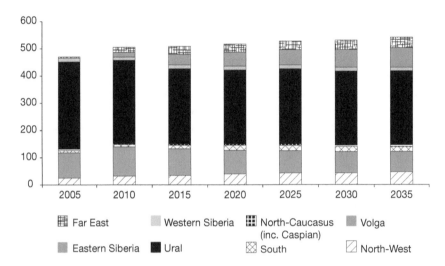

Figure 3.3   Oil production by region 2005–2035, in mtoe

Source: Global and Russian Energy Outlook until 2035. ERI RAS-REA. Moscow, 2012.

According to the ES-2030, production is to grow more slowly and export capacity additions will be limited: Russian oil output is projected to reach 510 mt in 2020 and up to 535 mt by 2030 – much lower than previous forecasts for 2030. Such a stagnation in oil production is not something that the government desires, but avoiding it will be very difficult – even to provide for stable oil production is a challenge for the whole Russian oil industry, with production from older fields declining rapidly.

Gas Industry

Today the Tyumen region in Western Siberia – specifically the Yamal Nenets Autonomous Okrug – is the source of 85 per cent of the country's gas production. In 2008 production here decreased significantly, with volumes expected to regain 2007 levels only in 2014. However, this decline was only temporary, as existing production and transportation infrastructure and low production costs are sufficient to secure the region's leading position as gas producer and main source of gas for export markets.

In the longer term, falling production from the 'Big Three' Russian gas fields (Urengoy, Yamburg, Medvezhye) will be replaced by production from new fields in the Tyumen region and the development of fields on the Yamal Peninsula, in East Siberia and in the Far East. By 2035, total Russian gas production is expected to have increased by 30–40 per cent thanks to the development of new fields in these remote areas, whereas output is expected to be halved in the currently developed regions (Stern, 2009; Makarov, Grigoriev and Mitrova, 2012: 155).

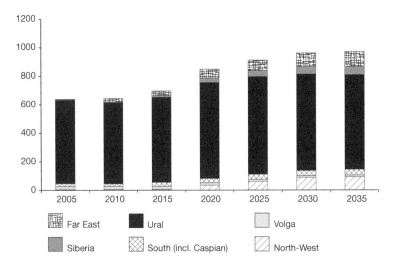

Figure 3.4   Gas production by region 2005–2035, in bcm

Source: Global and Russian Energy Outlook until 2035. ERI RAS-REA. Moscow, 2012

The dramatic decline in domestic and external demand has allowed Russia to be in less of a hurry to implement its mega-projects. Production and transportation costs for the new remote production areas will be much higher than in the Tyumen region, so projects there will make sense only if large volumes can be produced and exported (economy of scale). It thus seems reasonable to expect Russia to postpone realization of plans there until security of demand has been ensured. In the short- to medium-term perspective and depending on the market situation, production goals may be revised upwards or downwards. Russia has sufficient capacity and flexibility to meet higher demand during economic recovery. In the future, it will have to adjust its production volumes to the needs of the external market. Theoretically Russia's biggest gas upstream projects have a huge potential for flexibility, but without timely investments it will be very difficult to utilize this physical potential.

The Yamal Peninsula is the most important among these upstream project areas. Gazprom began laying the highly challenging Bovanenkovo–Ukhta gas pipeline in 2008, and production of the first gas started in October 2012. According to the General Scheme of Gas Sector Development, Yamal output is slated to peak in 2020 with production of between 135 and 175 bcm. In September 2009, Putin invited international oil companies to participate in Yamal LNG together with Gazprom and Novatek. Foreign companies were offered the opportunity to participate in the construction of an LNG plant, and in 2011 Total decided to join forces with Novatek in an ambitious Yamal-LNG project. This is unquestionably a challenging project: but the point here is that, for the first time, foreign companies were given an opportunity to get into Yamal, an area traditionally regarded as Gazprom's fiefdom. This might be the first sign of changing approaches towards upstream development in the new provinces. It could be extremely important for East Siberian development with its LNG projects, complicated gas content demanding gas processing plants construction – the participation of foreign investors appears crucial if these projects are to be realized.

Eastern Siberia and Far East have a potential to become a new centre of gas industry growth in Russia. Asia-Pacific gas demand is expected to more than double by 2030; there are forecasts of Chinese gas demand growth of 300 per cent by 2030, up to 330 bcm (IEA, 2007) – Russian gas could be an attractive long-term solution. In 2006, a Memorandum of Understanding was signed, announcing two directions of gas supply to China starting from 2011/2012. The western route – the Altai gas pipeline, total length 2700 km – is to be constructed on Russian territory to transport gas from Western Siberia. The eastern route is to provide gas from Eastern Siberia and entails construction of the Yakutia–Khabarovsk–Vladivostok pipeline, which Gazprom had scheduled for 2011. After the summit with China in June 2009, negotiations were resumed and in October 2009 a new MoU was signed, with supplies to start in 2014/2015. The price issue has not been solved yet, but China agreed to link the price of gas to the price of JCC ('Japanese Crude Cocktail'). The decision of the Chinese government to raise domestic prices and to link gas import prices to oil could make the projects profitable. An agreement on the gas price would result in rapid development of pipeline gas supplies to China (reaching up to 50–85 bcm by 2030 according to official plans) but the two parties have not been able to reach agreement; in the meantime the Chinese side has embarked on developing its own deposits of non-conventional gas.

The most attractive and practical Asian markets – Japan and South Korea – are LNG-focused. Japan's nuclear tragedy in Fukushima in 2011 made it clear that the region will need huge additional LNG supplies. Gazprom has speeded up development of the Sakhalin-3 project in addition to the Sakhalin-2 and has started work on the Vladivostok-LNG feasibility study.

As noted above, the second priority of Russia's external energy strategy is to diversify the product structure of exports. According to ES-2030, the share of LNG in gas export should reach 15 per cent by the year 2030. This goal can be achieved through the flexible development of LNG production in the Atlantic basin (Yamal LNG and Shtokman in the longer term) and by expanding LNG production in the Pacific basin with possible downstream development in Hong Kong and Taiwan, combined with active Russian participation in trading and arbitrage.

Nevertheless Russian gas export to Asia faces several new challenges after the economic crisis. Increasing competition with both domestic sources (shale gas in China) and alternative suppliers (LNG from Australia, the USA, Canada, the Middle East and possibly from East Africa as well as growing pipeline gas export from Central Asian states) creates a much more challenging environment for Russia's new and expensive projects in Eastern Siberia and Far East. Moreover, timing is becoming critical, as the window of opportunities might close and uncontracted market niches in these markets might be covered by non-Russian supplies. All these considerations are making Russia more hurried, as several investment decisions have to be made: Sakhalin-2 expansion, Vladivostok LNG, development of the Chayanda and Kovykta fields – all these projects are now to be speeded up.

Russia exports the largest share (about 80 per cent) of its gas to Europe. Europe is, of course, well-placed as a natural market for gas from Russia: all existing export pipelines go to Europe. On the other hand, Europe is already a mature market, approaching plateau demand in the foreseeable future. Here Russia aims to retain its current market share. Pricing is becoming a critical issue. Russia pays particular attention to protection of the existing long-term take-or-pay contracts, as they ensure the main export volumes and stability of the gas market, and are thus crucial to the realization of its mega-projects (Mitrova, 2011c).

Gas consumers in Europe have claimed that traditional netback market pricing based largely on oil products is now no longer logical because gas and oil product markets have diverged and there is very little switching capacity left. Recession and gas oversupply have speeded up this process. Price review pressure increases as consumers insist on modification of the contract system and changing the price terms. A gap between the prices set in long-term contracts and spot prices led buyers to shift to the spot markets – but this also became the reason for the beginning of the process of renegotiating contract provisions. In 2009, many large consumers succeeded in obtaining discounts or temporary renegotiation of contract provisions with a growing spot component in the price formulas. The development of gas spot markets spurs consumers to avoid binding contract prices in the wholesale long-term contracts with oil product indexes, European companies have insisted on modernization of existing contracts and a move to hub-based prices.

This is an absolutely new situation for Russia. Gazprom and the government have now to choose which targets are more important – export level or price level – and build the whole strategy accordingly. The question of pricing and contract transformation is central in this discussion.

Gazprom has continued to insist that oil-linked pricing must stay – although it has changed its position and now concedes that hub-based pricing can co-exist with oil-linked prices. But problem for Russian gas is not price formation, it is price *level*. As Russian gas is relatively a high-cost commodity – and will become even more costly if Arctic resources are to be developed in future – then gas prices significantly lower than the current spot level are potentially damaging. Depending on the strategy chosen, Russian gas export objectives should focus either on starting a smooth transition to hub-based pricing – continuing to resist this transition, fighting arbitrations ending in settlement or contract termination and continuing to export less during this period, but prolonging the oil linkage for several years. In any case, the European market seems set to remain the focus of Russian gas export strategy, although it will become more difficult and competitive.

Security of Russian gas supply to Europe is a special issue. Russian gas is delivered to Europe through a grid of transit pipelines. There is Brotherhood, running via the territory of Ukraine; Blue Stream, passing through the eastern tip of Ukraine and from near Tuapse to Samsun in Turkey; Soyuz, running through Ukraine, with a trans-Balkan line going via Moldova to Romania to Bulgaria and to

Turkey; Northern Lights, running via Belarus and Ukraine; and the Yamal–Europe pipeline through Belarus. Only the Yamal–Europe pipeline circumvents Ukrainian territory. Until recently, 80 per cent of Russian gas supplies to Europe went through Ukraine: and Ukraine cannot be replaced as a transit country even if new projects are realized. About 20 per cent of Russian gas export to the EU goes through Belarus (35 bcm). The reliability of this transit route became an acute issue after the 2004 Russo–Belorussian conflict, and in the wake of the crisis Russia managed to gain better control over the Belarusian pipeline system (RIA Novosti, 2011).

Relations with transit countries are crucial for the sustainability of Russian gas exports. In 2009 Russia and the EU signed a memorandum on an early warning mechanism in the oil, gas and electricity sphere, to deal with possible interruptions in supplies of energy resources to Europe. The memorandum specifies step-by-step measures to be implemented in order to prevent a possible conflict in the supply of energy resources. Otherwise, the memorandum is not legally binding and is limited to confirming the goodwill of both sides to collaborate on the basis of mutual respect and benefit. However, given the experience of gas conflicts in 2006/2007 and 2009, when letters and faxes from the Russian side were delivered to EU offices (which had closed for Christmas holidays), even this small progress is positive. It is quite understandable that the Russian Federation should pay special attention to the Nord Stream and South Stream projects, which bypass the territories of Ukraine and Belarus. But extremely weak European gas demand may put into question the economic efficiency and utilization rates of these projects, which is also an expected post-crisis development. With external gas markets generally becoming less certain and more competitive, Russia will have to make a number of strategic choices and decisions in order to protect and expand its market positions.

Coal Industry

During the crisis, coal mining fell in reaction to reduced domestic and foreign demand. This sector is extremely sensitive to the external environment, since a huge proportion of Russian coal is exported. In the inertia scenario foreseen in the ES-2030, coal output virtually stabilizes soon after 2010. By contrast, the innovation scenario, with high growth rates of non-gas power generation and faster development of Russia's Eastern regions (which have abundant resources of local cheap coal), the coal industry will be booming. (On the current problems of the Russian coal industry, see Markova and Churashev, 2012.)

Today the region with the highest production of coal is Kuzbass. Coal production is planned to increase substantially in the Kuznetsk (Kuzbas) and Kansk-Achinsk coal basins. In the medium to long term, coal production will also increase in the new fields in Eastern Siberia and Far East (Urgalskoe, Elegest, Elga, Apsatskoe). In addition, once their economic feasibility has been confirmed, other fields, like Seidi (Komi Republic) and Sosva (Khanty-Mansi Autonomous Okrug – Ugra) and the Beringovsky Coal Basin (Chukotka Autonomous District) will start producing coal.

Further development of the export potential of the Russian coal industry will necessitate the construction of the new export infrastructure – first of all port infrastructure (port East, Vanino, Ust–Luga, Murmansk deepwater port) and new ports with coal terminals, including on the Black Sea and Pacific coasts. As this sector is currently dominated by private companies and the government has rather limited involvement in the development of the coal industry, future international cooperation with foreign companies will depend mainly on the situation on external markets and the profitability of such cooperation for Russian and foreign companies alike.

However, in 2010 the Russian Ministry of Energy decided to prepare an official strategy for the long-term development of the country's coal industry until 2030, finally approved by the Russian government on 24 January 2012. This document outlines the main challenges faced by the coal sector and proposes steps to be taken to improve the situation and to make it better prepared to cope with changing market conditions (Ministry of Energy, 2010). However, it remains to be seen how this governmental strategy will be implemented in a sector where privately-owned companies have a much stronger position than in oil or gas, but the fact that government decided to approve this document indicates the strategic importance of coal in Russia. There are at least two good reasons why the Russian authorities should pay attention to the development of country's coal resources: first, the interest in increasing the share of coal in power generation, so as to make more gas available for export; second, the abundance of coal in Russia and coal's 'strategic return' as the future energy commodity expected to surpass oil as the world's top energy resource by 2017.[1]

Conclusions

The recent global economic crisis has had a massive impact on Russia's energy sector, with significant decreases in production and investments in response to lower demand and prices. The crisis has completely reconfigured the world energy market, making Russia's energy resources no longer solely a strategic asset, but also a curse. For the first time ever, Russia finds itself facing weak demand on the external markets, which limits its exports.

Finding a new place in the changing world is no easy task. The external environment for the Russian energy industry is changing dramatically, necessitating new and more flexible approaches. On the one hand, the slowdown in domestic and external demand can allow Russia to be in less of a hurry in developing new mega-projects, and has created additional incentives for cautious assessment of their economic feasibility in the new market environment. On the other hand, given the role of energy resource exports in budget formation and GDP growth, it is understandable that this situation looks so challenging for the Russian economy, which had become used to basing its growth solely on the export of hydrocarbons.

---

1   http://www.iea.org/newsroomandevents/pressreleases/2012/december/name,34908,en.html.

Russia has responded to the crisis by attempting to open up new markets and boosting the security of older ones by further developing its transit network (Makarov, Mitrova, Kulagin, 2011). The booming markets of Asia, China in particular, have become major priorities of Russia's state export strategy, and evident progress has been made in this field.

Russia has the potential to supply more gas to the global market. Not only Gazprom but also vertically integrated oil companies and independents have a huge potential for increased production – but only if there is significant growth in demand on the external markets. Access to capital and guarantees on the demand side are crucial for transforming this potential into actual additional supply volumes. The same picture can be observed in the coal industry: export is the main destination of all incremental coal production in Russia. With oil exports, however, the picture is quite different: these are expected to stagnate in any case, as the limited increases in production will be mainly absorbed by the growing domestic and Asia-Pacific region markets (Government of Russia 2009, Appendix 1).

New technologies, advanced project management and consistent regulatory framework are necessary to provide competitiveness for future Russian energy export. This is creating a new agenda for the country's decision makers, shifting their focus from volume maximization to competitiveness – and, it is hoped, marking new era in the energy policy of the Russian Federation.

References

Aalto, P. 2007. *The EU–Russian Energy Dialogue*. Farnham: Ashgate.

Åslund, A. 2006. Russia's energy policy: a framing comment. *Eurasian Geography and Economics*, 47(3), 321–8.

Astrov, V. 2010. Current state and prospects of the Russian energy sector. *Research Report* 363, Vienna Institute for International Economic Studies.

Averre, D. 2010. The EU, Russia and the shared neighbourhood: security, governance and energy. *European Security*, 19(4), 531–4.

Fernandez, R. 2009a. Russian gas exports have potential to grow through 2020. *Energy Policy*, 37(10), 4029–37.

Fernandez, R. 2009b. Some scenarios for Russian oil exports up to 2020. *Europe–Asia Studies*, 61(8), 1441–59.

Goldthau, A. 2008. Improving Russian energy efficiency: next steps. *Russian Analytical Digest*, 46, 25 September, 9–12.

Government of Russia 2008. *General'naya skhema razvitya gazovoy otrasli na period do 2030 goda (General Scheme of Gas Industry Development until 2030)* Available at: http://www.energyland.info/library-show-2928 [accessed 15 November 2012].

Government of Russia 2009. *Energeticheskaya strategiya Rossii na period do 2030 goda [Energy Strategy of Russia through 2030]*. (Decree of the Government of RF # 1715-p). Moscow: Government of the Russian Federation.

Grigoryev, L. 2009. Mirovaya retsessiya i energeticheskie rynki, in S. Chebanov (ed.), *Mirovoy krizis i global'nyye perspektivy energeticheskikh rynkov.* Moscow: IMEMO RAN, 39–56.

IEA 2007. *World Energy Outlook.* Available at: http://www.iea.org/weo/2007.asp [accessed 15 November 2012].

Makarov, A., F. Veselov, T. Mitrova and Y. Kulagin 2011. Russian scenarios of energy production, demand and export beyond 2030 – extending the Energy Strategy horizon, in *Energy Scenarios and Forecasts: Role of Natural Gas and EU–Russia Energy Cooperation to 2050.* Moscow: ERIRAS.

Makarov, A., L. Grigoriev and T. Mitrova 2012. *Global and Russian Energy Outlook until 2035.* Available at: http://www.eriras.ru/files/Outlook_2012_eng_light.pdf [accessed 15 November 2012].

Markova, V. and V. Churashev 2012. Operet'sya na zapad i idti na vostok, 24 December]. Available at: http://expert.ru/siberia/2013/01/operetsya-na-zapad-i-idti-na-vostok/ [accessed 15 January 2013].

Ministry of Energy 2010. *Dolgosrochnaya programma razvitiya ugol'noj promyshlennosti Rossii na period do 2030 goda.* Available at www.rg.ru/pril/64/07/80/14_ghu.doc [accessed 15 January 2013].

Mitrova, T. 2009. Energy markets in a turbulent zone. *Russia in Global Affairs*, 3, July–September, 58–67. Available at: http://eng.globalaffairs.ru/number/n_13586 [accessed 15 November 2012].

Mitrova, T. 2011a. *Evolutsiya rynkov prirodnogo gaza [Evolution of petroleum market].* Saarbrücken: Lambert Academic Publishing.

Mitrova, T. 2011b. Long-term prospects of the global energy markets: the view of an energy supplier. Available at: http://www.eriras.ru/images/papers/prague.ppt [accessed 15 November 2012].

Mitrova, T. 2011c. Russia & the global gas market. How can Russia be proactive in this changing market environment & how will this impact the global gas market. In *FLAME 2011.* Amsterdam.

Mitrova, T. 2011d. Strategy of the Russian energy sector development with its implication for the technologies. Available at: http://www.eriras.ru/papers/mitrova_japan.pdf [accessed 15 November 2012].

RIA Novosti 2011. *Lukashenko says Beltransgaz sale to Gazprom 'very profitable'.* Available at: http://en.rian.ru/world/20111223/170444388.html [accessed 15 November 2012].

Romanova, T. 2010. Energy security without panic: Russia–EU energy dialogue moving back to economy. *Russia in Global Affairs*, 9(2). Available at: http://eng.globalaffairs.ru/number/Energy_Security_Without_Panic-14900 [accessed 15 November 2012].

Sergeyev, P. 2010a. Neftegazovyy kompleks Rossii: problemy razvitiya i diversifikatsii [Petroleum issues of Russia: problems of development in diversification], in *Mineralno-syrevyye resursy i ekonomicheskoye razvitye. Sbornik statey*, edited by V. Kondratiev and Y. Adno, 136–43. Moscow: IMEMO.

Sergeyev, P. 2010b. Rossiya v mirovom energosnabzhenii [Russian in world energy provision], in *Mineralno-syrevyye resursy i ekonomicheskoye razvitye. Sbornik statey*, edited by V. Kondratiev and Y. Adno. Moscow: IMEMO, 80–96.

Shadrina, E. 2010. Russia's foreign energy policy: norms, ideas and driving dynamics. Electronic Publications of Pan-European Institute (18) 2010. Available at: http://www.tse.fi/FI/yksikot/erillislaitokset/pei/Documents/Julkaisut/Shadrina_final_netti.pdf [accessed 15 November 2012].

Sherr, J. 2010. The Russia–EU energy relationship: getting it right. *The International Spectator*, 45(2), 55–68.

Smirnov, V., D. Yanchenko and S. Reshetilo 2011. Cena Kilowata. Mirovaya konkurentnosposobnost rossiyskikh firm obuslovlena nizkimi energotarifami [Price of a kilowatt. World competiteveness of Russian companies is explained by low prices]. *Nezavisimaya Gazeta*, 13 April.

Stern, J. P. 2009. The Russian gas balance 2015: difficult years ahead, in *Russian and CIS Gas Market and their Impact on Europe*, edited by S. Pirani, Oxford: Oxford University Press, 54–92.

Strukova, V. 2011. Metody stimulirovaniya vozobnovlyayemoy energii: vozmozhnosti dlya Rossii. Available at: http://www.eriras.ru/images/papers/mehanizmi.ppt [accessed 15 November 2012].

Tabata, S. 2009. The influence of high oil prices on the Russian economy: a comparison with Saudi Arabia. *Eurasian Geography and Economics*, 50(1), 75–92.

Tarazanov, I. 2009a. Itogi raboty ugol'noj promyshlennosti Rossii za yanvar'-mart 2009 goda [Russian coal industry in January – March 2009]. Available at http://www.ugolinfo.ru/itogi2009q1.html [accessed 15 November 2012].

Tarazanov, I. 2009b. Itogi raboty ugol'noj promyshlennosti Rossii za 2009 god [Russian coal industry in 2009]. Available at http://www.ugolinfo.ru/itogi2009all.html [accessed 15 November 2012].

Zagashvili, V. 2010. Rol i mesto Rossii na mirovom rynke syrya [Role and stance of Russia in world market of raw materials] in *Mineralno-syrevyye resursy i ekonomicheskoye razvitye. Sbornik statey*, edited by V. Kondratiev and Y. Adno. Moscow: IMEMO, 31–52.

Chapter 4

The Modernization Debate and Energy: Is Russia an 'Energy Superpower'?

Derek Averre[1]

Introduction

Two crises faced by Russia in 2008 – the conflict with Georgia in August of that year and the global economic and financial shock – have had a marked impact on the country's external relations. Much attention has focused on political-military issues. In spite of the more favourable external environment created by the Obama-inspired 'reset', largely replicated in Europe in a more sober and pragmatic approach to Russia, Moscow is still perceived as challenging the West and aiming to carve out, in the words of then President Dmitrii Medvedev, a 'traditional sphere of interests'[2] in its neighbourhood. As seen from Moscow, differences over key aspects of security governance in Europe and its institutional architecture have not been properly addressed: there is a need to revisit, and in some instances revise, the current arrangements. This central narrative in Russian foreign policy was crystallized in Medvedev's proposals for a legally binding European Security Treaty. A widespread assumption is that Moscow is prepared to use the country's vast hydrocarbons reserves, increased prices for which contributed substantially to its economic resurgence prior to 2008, as a political lever in support of its supposedly 'geopolitical' ambitions to cement its 'zone of influence' in its neighbourhood, and to divide and rule in Europe (Paillard, 2010).[3]

However, a second narrative, that of 'modernization', emerged during the Medvedev presidency, spurred by the shock of the global economic and financial crisis. As leading political economists have pointed out, the crisis underscored Russia's interdependence with global structures: 'many in Russia thought that they were decoupled from what happened in the leading industrial economies ... some observers have argued that Russia's crisis is home-grown. It is not. Russia's structures and policies have determined how the shock played out in Russia, but

---

1   The author would like to acknowledge helpful comments made on a previous draft of this paper by Richard Connolly, CREES, University of Birmingham.

2   Transcript of meeting with the participants in the International Club Valdai, 12 September 2008, available at: http://archive.kremlin.ru/eng/speeches/2008/09/12/1644_type82912type82917type84779_206409.shtml (accessed: 11 April 2011).

3   For a more subtle interpretation see Kaveshnikov (2010).

the shock itself was external' (Gaddy and Ickes, 2010: 281; see also Hanson, 2009). This crisis thus has far-reaching implications for Russia's external relations. Medvedev himself emphasized this point, linking the conflict with Georgia, the economic shock and the launch of the modernization programme as major factors shaping Russia's foreign policy.[4]

In fact, both narratives – security and modernization – focus on the perceived need for a broader 'transformational ... project of economic modernization, social renewal, a new Euro-Atlantic security framework from Vancouver to Vladivostok and reforms of global governance' (Pabst, 2010). In a situation where energy resources will be crucial in Russia's modernization, the challenge for Russia's governing elite is – in the words of an industry source – to develop the sector and create 'a platform for the modernization of the entire economy'.[5] Thus the state's influence over the hydrocarbons sector, the use of regulatory instruments to establish control over assets, the distribution (or redistribution) of privileges, and the constraints placed on overseas investors will be key indicators of progress in the country's modernization as a whole (see Gustafson, 2012a: 382–411).

This contribution to the present volume aims to raise some fundamental questions about the role of energy in Russia's modernization: what kind of modernizing state the governing elite is trying to fashion, what is the influence of domestic interest groups over governance in the energy sector, and how this impacts on Russia's external relations. It first examines the motives underpinning the launch of the modernization programme, highlighting the challenges this poses to relations with its key energy market, Europe; second, it investigates competing ideas about how oil and gas resources should be used to stimulate Russia's modernization; and finally, it focuses on the key problem of rent addiction in Russia's economy. The central question arises: is modernization under Putin, now in his third term as President, capable of driving genuine change, and will current policies allow energy to become the 'locomotive' of Russia's national development?

Modernization: Russia's Domestic Challenges

The challenges of modernization in Russia are substantial and multifaceted. As one authoritative commentator has argued, a 'monumental' effort is needed in order 'to upgrade its economic clout, technological prowess, and societal appeal before it can claim the status of a world-class power' (Trenin, 2009). In launching the modernization programme Medvedev explicitly and repeatedly emphasized the

---

4   Speech at meeting with Russian ambassadors and permanent representatives in international organizations, 12 July 2010, available at: http://eng.news.kremlin.ru/transcripts/610/print (accessed: 23 September 2010).

5   Building on Success: Modernization and the Russian Oil and gas Sector, available at   http://www.bp.com/liveassets/bp_internet/russia/bp_russia_english/STAGING/local_assets/downloads_pdfs/s/CIS_Oil_Gas_Jan_2011_eng.pdf (accessed: 15 October 2012).

need, not only for technological advancement, but also to 'give impetus to progress in all areas, based on democratic values; to raise a generation of free, well-educated and creative people; to raise the living standards of our people to a fundamentally new level; to confirm Russia's status as a modern world power whose success is based on innovation' (Medvedev, 2010). He also asserted that Russia must establish 'a completely different legal culture, a respected court and efficient law enforcement agencies'[6] and 'overcome paternalism and parasitical attitudes'.[7]

The need for an effective foreign policy as a resource to underpin Russia's modernization in the context of an interdependent global economy was explicit in official pronouncements at the time (Medvedev, 2009). A draft foreign ministry document reinforced the point that Russia's foreign policy aims should be 'in close coordination with the needs of the country's complex modernization' and directed towards strengthened relations with Europe and the USA based on 'interpenetration of economies and cultures … the creation of "alliances for modernization"' (Lavrov, 2010a). Lavrov highlighted the importance of such partnerships in terms of global trends: 'The world is experiencing a turning-point and seeking new paths and models of development. International relations are being cardinally transformed and their former confrontational paradigm, reflected in zero-sum game logic, is becoming a thing of the past … security and development are indivisible'.[8] One source described this as 'virtually a new foreign policy doctrine', citing a senior European official as saying that 'at all of the most important negotiations, Russian representatives have been talking only about how Europe may invest in Russia's modernization' (*Newsweek Russia*, 2010).

The most visible policy outcome of Medvedev's initiative was the launch of a 'Partnership for Modernization' (PfM) with the European Union, Russia's main trading partner, at the November 2009 EU–Russia summit. This has since been supplemented by individual PfMs with almost all of the EU member states. At the December 2010 summit, the president of the European Council, Herman van Rompuy, declared that Russia's modernization should be based on 'democratic values, by building a modern economy, and by encouraging the active involvement of civil society'. This was backed up by Commission President Jose Manuel Barroso's call for the establishment of the rule of law in Russia, emphasizing the link with negotiations on a new EU–Russia legal agreement which should contain 'substantive provisions in all key areas, including trade, investment and energy'.[9]

---

6 Interview with CNN, 20 September 2009, available at: http://eng.kremlin.ru/news/1622.

7 Transcript of meeting of Commission for Modernization and Technological Development of the RF Economy, 11 February 2010, available at: http://news.kremlin.ru/transcripts/6844/print (accessed: 8 March 2010).

8 http://www.mid.ru/brp_4.nsf/0/F3DEB8AB978ABD45C3257728005604D1 19 May 2010 (accessed: 20 May 2010).

9 Press statements following EU–Russia Summit, 7 December 2010, available at: http://eng.news.kremlin.ru/transcripts/1435/print (accessed: 20 January 2011).

These statements chimed with Medvedev's explicit emphasis on the link between economic and political transformation and social renewal: 'We need new standards of governance and public services, a higher quality of courts and law enforcement ... we need to make day-to-day interaction between the state and its citizens transparent, clear and simple. The understanding that government officials serve the people rather than control their lives is the foundation of democracy' (Medvedev, 2010).

However, this apparent agreement on fundamental principles was offset by Medvedev's rejection – often expressed in the very same speeches and articles promoting modernization – of a return to what is widely perceived by Russian elites and society as the 'chaos' of oligarchic capitalism in the 1990s, and a refusal to copy foreign models of development in favour of a gradual approach to political and social change (Medvedev, 2009). The belief – undoubtedly shared by Putin as the country's preeminent political figure – that, while Russia's political system requires institutional improvement, it can lay claim to a measure of stable democracy, coexists with the notion that 'our people are not prepared to accept a fully-fledged democracy in the true sense of the word. They are not ready to fully experience democracy and gain a sense of involvement in and responsibility for the political processes'.[10]

This belief in 'evolutionary' democracy specific to Russia underpins the world view of the country's conservative political elites, who perceive a global shift towards plurality and competition in terms of development models (see Zevelev, 2009, Lukin, 2009, Averre, 2008).[11] Russia's governing elite interprets the global financial-economic crisis as 'the collapse of the sociocultural order which came together under the leading role of the West, its values and practice over recent centuries', and argues that the sovereignty of strong, independent states is the bedrock of the international system (Lavrov, 2009). As Lavrov has stated:

> Of course, we have taken on fundamental democratic values. We will uphold them, primarily for our own internal development, but we will do it primarily with due regard for our history, our culture, our traditions and with due regard for how we ourselves, our people want to see ourselves and our country in this world.[12]

---

10   Meeting with leading Russian and foreign political analysts, 10 September 2010, transcript available at: http://eng.news.kremlin.ru/transcripts/919/print (accessed: 27 October 2010).

11   Pabst (2010) argues that this is a feature of the BRIC countries as a whole, whose 'demand for more democratic global governance is at odds with their strong stand in support of state sovereignty ... they (rightly) oppose "homogenizing, liberal social and economic models" and defend every nation's right to choose its own path towards development and modernization; hence the emphasis on state power in the economy at the national and global levels'.

12   Interview with TV channel *Russia Today*, 16 October 2009, available at: http://www.mid.ru/brp_4.nsf/0/D4DB1B77CA438C40C32576550030DFBC, 20 October 2009 (accessed: 10 November 2009).

Authoritative commentators have thus concluded that 'conservative modernization ... has become the *Leitmotif* of the Kremlin's policy agenda' (Trenin, 2010, see also Malinova, 2012). This affects Russia's relationship with the EU: 'Now, if you read what President Medvedev is writing and saying, there are no words about integration, there are rather cooperation and interaction but without common values, without institutions. There is a modern partnership for modernization based on Russia picking up what it needs from the West, and that's it'.[13] The PfM – though ambitious in its declared scope and aims – is vague, without a definitive list of priorities, and suffers from the lack of trust and dearth of political will that has characterized the relationship to date (Entin, 2010). A leading figure in the European Parliament has criticized Moscow for 'trying to downgrade the new successor agreement to the Partnership and Cooperation Agreement – the legal basis of EU–Russia relations – to the level of a framework document' in preference to the Partnership for Modernization 'which was promoted as a politically comprehensive but legally non-binding instrument' (Severin, 2010). According to Sergei Karaganov, chair of the presidium of the influential Russian Council on Foreign and Defence Policy,

> the values gap [with Europe] on the Russian side is growing. The national state ... and sovereignty is the main thing which moves the Russian elite at the present stage of its development. This means a refusal to build a democratic society and a progressive deliberalization of the country, a refusal of law, of seeking support from the law ... the Russian elite and Russian society have for the moment repudiated freedom and democracy.[14]

The PfM appears less as a qualitative breakthrough in relations and more as an enhanced version of the sectoral dialogues going on within the EU–Russia Common Spaces format, which have been painfully slow and limited in their implementation and scope; indeed, it was stated at the Rostov summit that 'the sectoral dialogues will be a key implementation instrument for the Partnership for Modernization'.[15] As one Russian report concludes, despite progress in some of these dialogues, the two sides 'have no mutual understanding of what these common spaces should be' (Valdai Club, 2010). With regard to EU–Russia energy cooperation, a legal scholar has described it as based on "soft law' mechanisms, which certainly have their merits but cannot conceal the lack of legally binding norms regarding investment protection, transit or dispute resolution' (van Elsuwege, 2012).

---

13   Fedor Lukyanov; see *Russia's International Role in the Coming Decade* transcript of Chatham House event, 10 December 2009, available at: http://www.chathamhouse.org. uk/research/russia_eurasia/research_events/view/-/id/1378/ (accessed: 17 February 2010).

14   *Novyi geostrategicheskii vybor Rossii i Evropa* (*Russia's new geostrategic choice and Europe*), round table at State University Higher School of Economics, 12 November 2009.

15   See Joint Statement at Rostov summit 10546/10, 1 June 2010, available at: http:// www.consilium.europa.eu/uedocs/cms_Data/docs/pressdata/en/er/114747.pdf.

The Role of the Energy Sector in Modernization

The importance of Europe's markets to Russian producers and the EU's own interest in oil and gas trade keep energy relations at the forefront of EU–Russia relations. Moreover, Medvedev has identified the energy sector – along with biomedicine, space and information technologies, and communications – as a priority in Russia's modernization programme.[16] At a meeting of the RF Security Council, Medvedev identified four key areas which require attention: energy efficiency, the generation of energy from non-fuel sources such as hydropower, environmental safety in the energy sector and, as a key objective, modernization and transition to an innovative development model of the fuel and energy sector infrastructure.[17]

Despite the well-documented difficulties in relations between Moscow and Brussels – disputes over gas deals and the lack of mechanisms for interaction between the business communities of the two sides (Zhiznin, 2010) – some progress has been made on the energy dialogue within the EU–Russia PfM. The preparation of a 'road' map for cooperation to 2050 is ongoing, focusing on energy efficiency and renewable energy, with support for pilot projects and investment facilitation through cooperation with international financial institutions, as well as further improvements in the energy regulatory framework. Specific projects include cooperation on grid development plans and power systems interconnections; establishing a secure legal environment for investment through EU–Russia Energy Dialogue meetings; exploring the possibilities for an EU–Russia centre for energy efficiency (a project is being implemented within the Northern Dimension Environmental Partnership); developing new options for gas, including in the transport sector; and encouraging specific industry projects.[18] Much of Russia's focus is on the demand side of the energy equation, with less focus on supply-side issues of fiscal regulation, and investment and property rights. In declarative terms at least, however, there is still emphasis in the PfM as a whole on improving and harmonizing the regulatory framework for business and creating 'joint projects in the area of the rule of law'.[19]

---

16   Speech at meeting with Russian ambassadors and permanent representatives in international organizations, 12 July 2010 at http://eng.kremlin.ru/transcripts/610.

17   *Security Council meeting on the current state of and measures to guarantee Russia's energy security*, 13 December 2010, available at: http://eng.news.kremlin.ru/transcripts/1468/print (accessed: 20 January 2011).

18   Progress report agreed by the coordinators of the EU–Russia Partnership for Modernisation for information to the EU–Russia Summit of 21 December 2012, available at: http://eeas.europa.eu/delegations/russia/documents/eu_russia/p4mdec2012_en.pdf (accessed: 8 March 2013). See also Work plan for activities within the EU-Russia Partnership for Modernisation, 23 May 2012, available at: http://eeas.europa.eu/delegations/russia/documents/eu_russia/p4m-wp-may2012_en.rtf (accessed: 8 March 2013).

19   *EU-Russia – taking the Partnership for Modernization forward*, 12 October 2012, available at http://eeas.europa.eu/delegations/russia/press_corner/all_news/news/2012/20121012_01_en.htm (accessed: 17 October 2012).

There is no doubt that partnerships with foreign investors in Europe, not to mention the US and Asia-Pacific countries, will be crucial in assisting Russian corporations to exploit dwindling resources in Western Siberia and the lower Volga, access more remote and difficult-to-access sources of hydrocarbons in Eastern Siberia and the Arctic, upgrade processing and pipeline facilities, and introduce clean energy and energy-saving technologies. The EU, in the words of one of its senior officials dealing with Moscow, 'is the natural partner for this diversification' away from Russia's resource export dependence'.[20] A detailed report by specialists at Russia's Institute of Contemporary Development (INSOR), the think-tank heading the modernization programme whose board of trustees is chaired by Medvedev himself, agrees with many of these conclusions. It adds that increasing the value of energy resources by processing primary feedstock to produce new materials very much in demand on the domestic and world markets – something that would require both government coordination and substantial capital investment, which could come from earnings from the energy sector itself – is one route forward (INSOR, 2010a).

Gustafson (2012b) has identified three fundamental competing visions for Russia's modernization. The first, associated more with Medvedev, promotes the idea of investing energy revenues in hi-tech manufacturing and innovation, exemplified in the founding of the Skolkovo Innovation Centre, and reducing consumption by becoming more energy efficient, in a Soviet-style 'great leap' towards modernization. The second is the approach of the liberal economist Aleksei Kudrin, Russia's finance minister from 2000 to 2011 and the main architect of Putin's fiscal and budgetary policy, who has warned that reliance on high volumes of energy production can not continue and has advocated improving the regulatory climate, and reducing the role of massive state corporations, to attract investment and kick-start the economy. This in turn contrasts with Putin's vision, which focuses on hydrocarbons as the locomotive of growth and the source of revenues to support other strategic sectors, including the defence complex, and envisages a larger role for state corporations in the energy business.

Prominent US political economists, while forecasting that hydrocarbons resources will remain at the hub of Russia's economy in the longer term, have questioned 'the conventional wisdom that diversification of Russia's economy (away from oil and gas) is a desirable objective that will render it less vulnerable to external shocks'; they argue that sharing the risk in exploiting energy sources with Western companies and investing in global assets would be a better strategy. They identify rent addiction, rather than resource dependence, as constituting the main challenge for the government (Gaddy and Ickes, 2010: 281, 292; Gustafson, 2012a: 4–5): a crucial factor in Russia's broader modernization that we turn to next.

---

20 *The EU-Russia Partnership for Modernisation – key to unfold the full potential of EU-Russia relations*, Gunnar Wiegand, Director, European External Action Service, EU-Russia Partnership for Modernisation conference, 12 October 2012 (available at http://eeas. europa.eu/delegations/russia/documents/news/20121008_speech_p4m_eng.doc (accessed: 17 October 2012).

Governance in the Energy Sector: Prospects for Genuine Modernization?

As outlined above, the link between economic transformation and political and social renewal has been a key idea in the modernization narrative from the outset. An immediate obstacle to change, widely recognized in the literature on its political economy, is the fusion of politics and big business in Russia, which economists have characterized as 'state-managed network capitalism [which] operates primarily on a cognitive institutional pillar, rather than the more formal, rule-based regulative pillar' and which is 'socially embedded in Russian culture' (Puffer and McCarthy, 2007: 3–4).

The close relationship between Russia's power structures and economic agents in the energy sector – what one expert calls the Russian bureaucracy and 'the hydrocarbons *nomenklatura*' (Arbatova, 2008; see also Krawatzek and Kefferputz, 2010) – allows these interest groups to control and slow down modernization, particularly the top–down variant which the government is currently promoting. There is an argument for state involvement in Russia's key industries; but whether the conservative and corporatist domestic policies currently supported by an influential part of Russia's governing elite can be transformed by the imperatives of modernization in the contemporary global environment is open to question. Igor Yurgens, chairman of INSOR, has argued that the economic crisis demonstrated that the Russian model of forced growth, based on energy exports and increased state involvement in the economy, has exhausted itself. Transition to balanced growth, with improvement of the legal and regulatory environment, is needed to mitigate the bureaucracy's transfer of rents from the energy sector to state investment programmes and government spending on welfare and subsidies (Yurgens, 2010; see also Gustafson, 2012b, Hanson, 2009). Kudrin himself has commented that energy 'from being a locomotive for the economy, has become a brake' (Gustafson, 2012a: 505).

While not subscribing to the simplistic idea that, in Russia, 'raw political power determines energy policy' (Paillard, 2010), the present writer agrees with the argument that addiction to resource (especially oil and gas) rents, and the system to manage these rents put in place by Putin and carried on by Medvedev during his tenure, are key to developments in the economy as a whole which means that changes in the political economy of Russia will necessarily lead to changes to the rent management system (Gaddy and Ickes, 2010: 282).

The governing elite is locked into this system of rent management just as much as those representing the corporate interests of the oil and gas companies. Governance in the energy sector thus lies at the heart of Russia's system, which can be characterized as a 'limited-access order' where ruling elites limit political, economic and social access to power, thereby enabling them to capture and distribute economic rents via informal and often arbitrary contractual arrangements (Connolly, 2012: 23–5). One Russian commentator has identified the 'closed circuits' of this system and argued that 'the rampant corruption of the past few decades does not signal any deviation from or the corrosion of the system – it itself has been transformed into a system-building component' (Badovsky, 2009, see also Greene and Trenin, 2010). The problem is compounded by the weak influence of non-state and non-corporate entities in Russian society. A former UK ambassador to Moscow, pointing to the absence of renewal within the ruling elite, argues that there has been 'no attempt to launch a systemic modernization agenda in Russia'; moreover, even if the governing elite were to propose such an agenda 'it is also questionable whether there would be popular understanding of, let alone support, for radical change' (Wood, 2009).[21] Other authoritative commentators contend that the 'nervous aversion to large-scale reforms as such and a growing gap between the official policy discourse and reality on the ground' can be traced back to centralized power which 'had the effect of cutting off feedback mechanisms' and reveals 'the deep structural weakness in the foundations on which twenty-first century Russian power has been built' (Greene and Trenin, 2009).[22]

The vision of a diversified and innovative economy, with a dynamic middle class, demands a different approach: diluting state control of the biggest sectors of the economy, ending restrictive measures, mitigating the influence of rent-seeking elites and creating a larger space for civil society in the political system (Trenin, 2010). Reducing reliance on hydrocarbons rents to sustain stability is an important part of this. A modernizing state could become part of the solution, but if the current system of bureaucratic capitalism persists, it will simply perpetuate the problem. Put simply, non-transparent decision-making structures, the fusion of power and wealth in the 'petro-state' and the legacy of 'sovereign' foreign policy thinking may combine to thwart systemic reform, despite recognition of the need for change in a rapidly transforming global environment (Lo and Shevtsova, 2012). The risk for Moscow is of a double failure, of domestic modernization and of its attempts to project the image of a modern and attractive world power.

At present it is not easy to discern where agents of renewal will emerge from. Change is most likely to be effected by those close to power; as one of the more perceptive Russian scholars has pointed out, 'the Russian ruling class is not homogeneous' (Arbatova, 2008). While liberal influence in the political elite is marginal, some do understand the urgency of renewal and modernization

---

21　Mankoff (2010) argues that 'the Kremlin's ability to reconcile the competing demands of the populace and the oligarchic elite with its own interest in preserving the political and economic status quo appears increasingly questionable ... All these elements of Russia's international resurgence have suffered in the course of the crisis, forcing Moscow to reconsider some of the basic principles underpinning its recent foreign policy'. Nefedova, Trevish and Pallot (2010: 214–15) analyse the deteriorating situation in Russia's regions and predict a '"middle road"' between building a '"civilized market"' and welfare state' on the one hand and 're-institution of state control' on the other – a 'quasi-market, corruption, and a grey economy, no doubt with some minor adjustments – in short, a continuation of the status quo'.

22　See also Arbatov (2007) who writes: 'The main objective today is not to strengthen the "vertical chain of command" but to establish effective control over it, making it more governable, and restoring feedback mechanisms between society and state'.

(see Hanson and Teague, 2013). As suggested earlier, this has been reflected in foreign policy narratives. Echoing Medvedev's earlier pronouncements, foreign minister Lavrov writes: 'The Russian leadership has made its priority investing in people ... the policy of complex modernization of the country is based on the values and institutions of democracy, a socially oriented market economy and the need to open up human potential to the absolute maximum' (Lavrov, 2010b). Again, however, this narrative exists side-by-side with a firm commitment to Russia's sovereignty, references to the 'crisis of the Western-centric international system' and question marks over 'the claim to universality of Western models of development and values systems' (MFA, 2010). As one leading Russian IR scholar argues, 'Russia is positioning itself as a different, alternative or even more genuine Europe'; current Russian policies – 'an amazing hybrid of modernization and restoration' – are 'part of the struggle for the existing ideological and political resources rather than the search for a radically different path' (Morozov, 2009).

Nevertheless, the tension between the narratives of sovereignty and modernization – both of them linked in the minds of Russia's governing elite with the pursuit of structural power advantages in the international system – arguably opens up space for progressive groups to challenge the more archaic features of the political, social and economic system. INSOR has led the argument for freeing up human capital and minimizing the top-down, bureaucratic-technocratic methods of modernization: 'The task is to create an economy that generates innovations rather than to generate innovations that are painfully implemented into an outdated economy' (INSOR, 2010b: 6). One commentator, while delivering a scathing assessment of the peculiar features of the corporatist system and the obstacles it places in the way of progress, argues that the 'generational shift' to 'the consumer and information society' marks the end of transition:

> the very awareness that the Soviet phase of history has ended, while the exploitation of the tapped-out economic, social and even political resources of the past is not possible anymore, may play the role of an important stimulus for accepting the idea of modernization and working out a relevant national project. In this sense, the upcoming years will not be post-Soviet anymore, as they will determine Russia's development for decades in the future' (Badovsky, 2009).

Indeed, this generational shift is something that Medvedev, in talking about the government's responsibility to change the Russian mindset, seems to be acutely aware of.[23]

---

23   See Meeting with leading Russian and foreign political analysts, 10 September 2010; speech at the Global Innovation Partnerships Forum, 11 October 2010, available at: http://eng.news.kremlin.ru/transcripts/1115/print (accessed: 27 October 2010). Kudrin has also voiced his belief in evolutionary change over the next generation (Hanson and Teague, 2013: 5).

The role of the Putin state, which claims to play the major part in effecting change, is likely to come under increasing scrutiny in Russia itself if the modernization programme fails to deliver growth. Richard Sakwa highlights the crisis-response nature of the modernization narrative and rightly questions the widely-accepted concept of the strong state in Russia:

> Contemporary Russia is faced not only by economic modernization tasks, the core of Medvedev's program, but also by the requirement to modernize the political system by increasing the effectiveness of existing institutions. This can only be achieved by reducing the informal powers of the administrative regime to allow more autonomy to public politics ... the deeper the crisis, the greater the emphasis on administrative governance, but this is more an indication of weakness than strength, and only exacerbates hybridity and, thus, crisis (Sakwa, 2010: 202).

In the absence of reform measures the spectre of managed modernization – a counterpart to the better known concept of 'managed democracy' – looms large.

Conclusions

The global economic crisis served to expose the shortcomings of Russia's rapid growth in the first few years of the last decade. This growth, largely fuelled by revenues from the hydrocarbons sector, has stalled as the leadership grapples with the dilemmas posed by modernization. State-dominated economic management, as well as an unwillingness or inability to loosen administrative shackles and harness the potential of business and society, generate the risk that modernization in Russia will be confined to a purely technocratic response to the economic crisis and will fail to open up the system to political and social renewal. In this case, the political pathologies and structural economic failings of a weak state may well simply be reproduced – a 'temporary surge' followed by an 'historic failure' (INSOR, 2010b: 6). Even Russian political commentators whose views are influential in official circles argue that

> it is in the long-term interests of Russia to reverse the trend towards curtailing democratic freedoms, which has obviously begun to deteriorate the efficiency of state governance, impede economic modernization due to the systemic proliferation of corruption, and worsen the country's positions in international competition. We believe that Russian society is interested in foreign assistance ... for improving the human rights situation and restraining the arbitrariness of uncontrolled bureaucracy'(Valdai Club, 2009:13).

This domestic situation impacts on external relations. Although the need for deeper political, economic and social engagement with the Western liberal democracies appeared initially to inspire the modernization narrative, many commentators discern the emergence of an opposite trend as technocratic Europeanization inspired by the illusion of the end of history did not open up any new intellectual horizons and Russia, at another spiral turn of its social transformation, has again failed to escape the role of an outsider and not-quite-European country (Morozov, 2009).

The role of the energy sector in Russia's development is of crucial importance, given both the overwhelming importance of oil and gas to the economy relative to other sectors and the fact that resource rent management has become built into the economic system; high levels of state spending rely on the government's ability to exploit hydrocarbons revenues. Despite the rhetorical commitment to modernization – diversification, innovation, the rule of law and development of human capital – this key problem remains to be tackled. Political economists are pessimistic: 'The problem for Russia is how to move away from addiction within the confines of the rent management system that Putin has created. There are no signs that Putin, or anyone else, has determined how to solve that problem' (Gaddy and Ickes, 2010: 308; Hanson and Teague, 2013: 4). Put simply, 'Medvedev's modernization "Viagra" was doomed from the very start and could not stimulate real change ... genuine post-industrial modernization of the economy needs a free individual, which means deep-rooted liberalization and the introduction of the rule of law and competition' (Lo and Shevtsova, 2012: 20).

At the time of writing there are few signs of a shift in the governing elite's adherence to conservative modernization reliant on energy rents. In the longer term, however, budgetary constraints and consequent pressure for social and political renewal – perhaps even before the expiry of the third Putin administration – may prompt calls for change. The ideas of liberal economists like Kudrin may act a catalyst for reform. While sustained efforts under initiatives such as the EU–Russia PfM are unlikely to effect a complete transformation of Russia, they may, through Russia's gradual convergence with European regulatory and legal standards, contribute to resolving contentious issues in energy and other areas of mutual interest, and ultimately to mitigating Moscow's emphasis on differing developmental models. Despite inevitable conflicts of interests, cooperation may provide the basis for strategic rather than short-term gains, and may yet assist materially in the modernization of Russia.

References

Arbatov, A. 2007. Bureaucracy on the rise. *Russia in Global Affairs* [online], 2. Available at: http://eng.globalaffairs.ru/number/n_8552 [accessed: 15 November 2012].

Arbatova, N. 2008. Rossiya posle prezidentskikh vyborov: vneshnepoliticheskie orientiry [Russia after the presidential elections: reference points in foreign policy], in *Mirovaya ekonomika i mezhdunarodnye otnosheniya*.

Averre, D. 2008. Russian foreign policy and the global political environment. *Problems of Post-Communism*, 55(5), 28–39.

Badovsky, D. 2009. Russia's modernization: at another fork in the road. *Russia in Global Affairs* [online], 3. Available at: http://eng.globalaffairs.ru/number/n_13584 [accessed: 15 November 2012].

Connolly, R. 2013. *The Economic Sources of Social Order Development in Post-Socialist Eastern Europe*. Routledge: Abingdon.

Entin, M. 2010. The partnership for modernisation – a way to bring Russia and the European Union together. *Internet-zhurnal 'Vsya Evropa'* [online], 9(47). Available at: http://www.mgimo.ru/alleurope/2006/47/article-mgimo-partnerstvo-3.html [accessed: 1 April 2011].

Gaddy, C.G. and B.W. Ickes 2010. Russia after the global financial crisis. *Eurasian Geography and Economics*, 51(3), 281–311.

Greene, S.A. and D. Trenin 2009. *Engaging Russia in an Era of Uncertainty*. Policy Brief: Carnegie Endowment for International Peace, no. 86, December.

Gustafson, T. 2012a. *Wheel of Fortune. The Battle for Oil and Power in Russia*. The Belknap Press of Harvard University Press, Cambridge Massachusetts and London.

Gustafson, T. 2012b. Putin's petroleum problem. *Russia in Global Affairs*, November/December.

Hanson, P. and E. Teague 2013. Liberal insiders and economic reform in Russia. Chatham House paper, REP2013/01, January.

Hanson, P. 2009. *Russia to 2020*. Finmeccanica Research Department Occasional Paper.

INSOR 2010a. Conditions and prospects for development of petrochemical and gas-derived chemicals industry in the Russian Federation [online: Institute of Contemporary Development]. Available at: http://www.riocenter.ru/files/Conditions%20and%20Prospects%20for%20Development%20of%20Petrochemical%20Industry%20in%20Russia.pdf [accessed: 1 April 2011].

INSOR 2010b. Russia in the 21st century: vision for the future [online: Institute of Contemporary Development]. Available at: http://www.insor-russia.ru/files/INSOR%20Russia%20in%20the%2021st%20century_ENG.pdf [accessed: 1 April 2011].

Kaveshnikov, N. 2010. The issue of energy security in relations between Russia and the European Union. *European Security*, 19(4), 585–605.

Krawatzek, F. and R. Kefferputz 2010. The same old modernisation game? Russian interpretations of modernisation [online: Centre for European Policy Studies working document]. Available at: http://aei.pitt.edu/15038/1/WD_337_Krawetzek_Kefferputz_Russian_Modernisation_latest.pdf [accessed: 1 April 2011].

Lavrov, S. 2009. Mezhdunarodye otnosheniya v novoi sisteme koordinat (International relations in a new system of coordinates). *Rossiiskaya gazeta*, 8 September.

Lavrov, S. 2010a. Programme for the effective use on a systemic basis of foreign policy factors aimed at the long-term development of the Russian Federation [online: The Daily Beast]. Available at: www.runewsweek.ru/country/34184/ [accessed: 28 May 2010].

Lavrov, S. 2010b. Odna na vsekh [One for all] [online: Itogi]. Available at: http://www.mid.ru/brp_4.nsf/0/882E7733C9E8F4B6C32577260020515F [accessed: 20 May 2010].

Lo, B. and L. Shevtsova 2012. A 21st century myth – authoritarian modernization in Russia and China [online: Carnegie Moscow Center]. Available at: http://www.carnegie.ru/en/pubs/books [accessed: 23 September 2012].

Lukin, A. 2009. Russia to reinforce the Asian vector. *Russia in Global Affairs*, 2.

Malinova, O. 2012. Yesche odin 'ryvok'? Obrazy kollektivnogo proshlogo, nastoyashchego i buduschego v sovremennykh diskussiyakh o modernizatii (One more 'leap'? Representations of collective past, present and future in the modern discussions about modernization). *Politicheskaya nauka*, 2, 49–72.

Mankoff, J. 2010. Internal and external impact of Russia's economic crisis, Russie.Nei.Visions. *IFRI* [online], 48. Available at: http://www.ifri.org/downloads/ifriengeconomiccrisisinrussiamankofffevrier2010.pdf [accessed: 23 September 2012].

Medvedev, D. 2009. *Go Russia!* Moscow: Kremlin. Available at: http://eng.kremlin.ru/news/298 [accessed: 2 November 2009].

Medvedev, D. 2010. Presidential address to the federal assembly of the Russian Federation [online: Kremlin News]. Available at: http://eng.news.kremlin.ru/transcripts/1384/print [accessed: 20 January 2011].

MFA 2010. Rossiiskaya diplomatiya v menyayushchemsya mire [Russian diplomacy in a changing world] [online: Federal'nyi spravochnik]. Available at: http://www.mid.ru/brp_4.nsf/0/120AA79468FE6685C325771500288220 [accessed: 20 May 2010].

Morozov, V. 2009. Policy transformation. *Russia in Global Affairs* [online], 4. Available at: http://eng.globalaffairs.ru/number/n_14242 [accessed: 20 May 2010].

Nefedova, T.G., A.I. Trevish and J. Pallot 2010. The 'crisis' geography of contemporary Russia. *Eurasian Geography and Economics*, 51(2), 203–17.

Newsweek Russia 2010. Pust' opyat' budet solntse [Let the sun shine again] [online: Newsweek Russia]. Available at: www.runewsweek.ru/country/34166/?print=y [accessed: 28 May 2010].

Pabst, A. 2010. Medvedev's 'third way': the unrealized potential. *Russia in Global Affairs*, 15 October.

Paillard, C-A. 2010. Russia and Europe's mutual energy dependence. *Journal of International Affairs*, 63(2), 103–24.

Puffer, S.M. and D.J. McCarthy 2007. Can Russia's state-managed, network capitalism be competitive? Institutional pull versus institutional push. *Journal of World Business*, 42(1), 1–13.

Sakwa, R. 2010. The dual state in Russia. *Post-Soviet Affairs*, 26(3), 185–206.

Severin, A. 2010. PCA and the modernisation partnership, EU–Russia centre report [online: EU-Russia Centre]. Available at: http://www.eu-russiacentre. org/wp-content/uploads/2008/10/EURC_review_XV_ENG.pdf [accessed: 28 May 2011].

Trenin, D. 2010. Russia's conservative modernization: a mission impossible? *The SAIS Review of International Affairs*, 30(1).

Trenin, D. 2009. Russia reborn. *Foreign Affairs*, 88(6), 64–78.

Valdai Club 2009. Towards a new Euro–Atlantic security architecture [online: Valdai International Discussion Club]. Available at: http://www.globalaffairs. ru/docs/Karaganov_eng.pdf [accessed: 10 September 2010].

Valdai Club 2010. K soyuzu Evropy [Towards an alliance of Europe] [online: Valdai International Discussion Club]. Available at: http://www.karaganov.ru/ docs/Karaganov_valdaj_rus.pdf [accessed: 29 November 2010].

Van Elsuwege, P. 2012. Towards a modernisation of EU-Russia legal relations architecture [online: CEURUS EU-Russia]. Available at: http://ceurus. ut.ee/wp-content/uploads/2011/06/EU-Russia-Paper-51.pdf [accessed: 30 November 2012].

Wood, A. 2009. *Russia's Coming Decade*. London: Chatham House.

Yurgens, I. 2010. The objectives and the price of modernisation in Russia. EU–Russia centre report [online: EU-Russia Centre]. Available at: http://www. eu-russiacentre.org/wp-content/uploads/2008/10/EURC_review_XV_ENG. pdf [accessed: 28 May 2011].

Zevelev, I. 2009. Russia's future: nation or civilization? *Russia in Global Affairs*, 4.

Zhiznin, S.Z. 2010. Ten years on. *Nezavisimaya gazeta*, 12 October.

Chapter 5

'Resource Curse' and Foreign Policy: Explaining Russia's Approach towards the EU

Irina Busygina and Mikhail Filippov

Introduction: Seeking Geopolitical Goals and Domestic Monopoly on Political Influence

With the rise in energy prices after the turn of the millennium, EU–Russia relations have begun to manifest a growing separation between the economic and political components: as EU reliance on trade with Russia has increased, political relations have deteriorated. Moreover, the question of energy security has become a highly contentious issue, and has contributed to redefining this relationship.

Russia's reputation in Europe as a reliable energy supplier had been built over decades of cooperation with the Soviet Union and pre-Putin Russia. The relationship developed gradually into one of mutual dependence. As two-thirds of Russia's energy exports go to the EU and Russia does not have – for the time being – other realistic and viable market options for selling great volumes of its natural gas, it must rely on European demand. Moreover, market prices for natural gas are much higher in Europe than in other potential markets. This is caused partly by the growing demand for energy, a demand that cannot be met by dwindling domestic production and must therefore be covered by external suppliers – and here Russia is by far the most important (Boussena and Locatelli, 2013; Victor, Jaffe, Hayes 2006).[1]

However, Russia found its reputation as a reliable supplier wiped out in the aftermath of the 'energy wars' between Russia and several former Soviet republics. Although some observers have argued that this aggressive energy tactic has been used 'exclusively within the CIS space' (Póti, 2008) and that energy deliveries to Europe were never manipulated, these conflicts not only damaged Russia's reputation but also led many European policy makers to view Russia as not merely a potential but an actual threat to European energy security. Russia, on the other hand, saw EU decisions and policies as a direct challenge to its dominant position on the European energy market and as an attempt at undermining the country's own strategic energy and geopolitical interests. This process culminated on 23 March 2009 when the Russian delegation, led by Minister of Energy Sergei Shmatko, walked out of the

---

1   'Russia is one of Europe's two main energy suppliers and is the only European supplier with large reserves of oil and natural gas and historically cheap prices'. See Stratfor 2013.

Brussels negotiations on the future of the Ukrainian pipeline system,[2] and Vladimir Putin threatened the EU with a revision of overall EU–Russia relations.[3]

In 2013 Gazprom has threatened Ukraine with a USD 7bn bill for buying less gas than agreed previously (Buckley, 2013). The Russian move was largely seen as retaliation for Ukraine's announcement of a deal with Royal Dutch Shell to develop shale gas in the eastern Ukrainian region of Donetsk. 'We do expect both parties, Russia and Ukraine, to ensure that there are no gas interruptions and gas supplies to flow normally as usual', said Marlene Holzner, spokesperson for EU Energy Commissioner Günther Oettinger.[4]

The question of spatial and infrastructural constraints in Russian energy policy and Russian attempts at addressing those issues at the strategic level are dealt with in greater detail in Pavel Baev's chapter in this volume, while the question of how the issue of energy pricing is influencing policy choices is taken up in Lunden's and Fjærtoft's chapter, where the relationship between domestic prices and export prices and the possible impacts of domestic price reform on availability of gas for export are discussed. In this chapter we offer a theory-based explanation of why cooperation between Russia and the EU has remained so severely limited in many areas of mutual interest, especially as regards policies in the post-Soviet region. Until very recently, EU–Russian relations were developing in a paradoxical way: as the dependence of both sides on mutual trade increased, their political relations deteriorated. Recent speculations as to a 'warming' of mutual relations could be explained either as result of the fall in oil and gas prices, or as caused by calls from the Russian elites for modernization and Russia's need for Western investments and know-how.

On the one hand, the Russian leadership has remained open to political dialogue and has contributed to the 'multiple small successes' of cooperation with the European Union (Smith and Webber, 2008: 83). On the other hand, Moscow now seems inclined not only to reject completely a path determined by Western democratic values, but to deny that such values even exist (Roberts, 2007; Semenenko, 2013). According to Morozov (2008), critical self-reflection within the EU has shifted into a feeling of superiority towards the neighbours, and this new EU identity conceptually challenges Russia's identity. In fact Russia is proposing its own version of the 'European idea' previously monopolized by the EU. On issues related to the post-Soviet space, Russia has increasingly acted in a conflictual manner, as if its interactions with the EU could produce only zero–sum outcomes there (Krastev, 2008; Mankoff, 2009; Sakwa, 2013).

In the following, we analyse the deterioration of the EU–Russian relations as a result of strategic choice. First of all, we connect the dynamics of the EU–Russia relations with the effect of high energy prices and the growing importance of energy trade (especially in natural gas) between Europe and Russia. The high share in Russia's national revenues from its exports of natural resources seems to create

---

2   http://lenta.ru/news/2009/03/23/gas/.

3   http://lenta.ru/news/2009/03/23/putin.

4   http://www.euractiv.com/energy/commission-hopes-best-new-russia-news-517410.

wrong incentives for economic development, growth, political system – and for foreign policy as well. Researchers have found an alarming degree of dependence between abundance of natural resources and stagnating economy and authoritarian rule. Literature exploring this dependence (called the 'resource curse') has studied why the states with abundant natural resources often suffer from 'bad governance' and weak democratic institutions. As a result, the concept of 'rentier states' has been elaborated – for those states that get from rent (generated from selling oil, gas or from foreign aid) a large share of state revenues; the recipient of rent is the government, which captures it. In a rentier state, the government does not depend upon making the economy work effectively: there are no incentives to create effective management structures. Resource rent not only allows the government to preserve a significant degree of autonomy, it also changes the very function of the state, from production and redistribution to distribution of resources (see Beblawi and Luciani, 1987; Karl, 1997; Wantchekon, 2002; Friedman, 2006).[5]

Second, we argue that Russian politicians have successfully used the deterioration of foreign relations as a mechanism to maintain the hybrid regime's status quo at home. Here we define 'hybrid regime' as one that combines elements of authoritarianism with electoral democracy and limited political competition, while at the same time profiting from economic ties with the West. The key premise is that the Russian leadership has domestic political incentives for sustaining a certain level of political conflict with the EU and many post-Soviet countries. Indeed, these incentives stem precisely from the importance of continued cooperation and openness on other issues. Moscow's increasingly conflictual relations with post-Soviet countries help the government to promote 'a virtual conflict' with the West over the post-Soviet area.

This 'virtual conflict', in turn, provides an opportunity to isolate domestic politics from international influences without having to resort to economic and informational isolation. It helps to combine both openness to the West (important for maintaining an open market economy), and effective silencing of any Western critics who might pose a danger to Russia's semi-democratic political system with its restricted political competition (Treisman, 2011). The domestic audience is made to believe that the West is a biased actor, whose views are compromised and whose communications are received but discredited and discounted. At the same time, this official anti-Westernism does not jeopardize the energy trade, which is driven by the West's demand for energy – including energy from Russia.

We will assume that, in its dealings with the EU and post-Soviet neighbours, the Russian government pursues a two-pronged set of objectives: traditional national geopolitical goals, and a need for conflict so as to create a domestic diversion. The diversionary conflict, in which the post-Soviet states feature as

---

5   However, recent comparative historical research conducted on the resource curse has shown that the link between resource rents and bad governance and undemocratic regimes is not so straightforward. See, for example, Arezki and Bruckner, 2011; Haber and Menaldo, 2011; Ross, 2012.

proxies of the West, provides an opportunity to isolate the Russian domestic audience from international influences. In consequence, Russian foreign policy has become founded on a counter-intuitive trade-off, as improved economic relations paradoxically allow for more, and not less, conflict in political relations with the West. The energy-trade dependence between EU and Russia is of key importance here, as it raises the stakes and paves the way for more political conflict with the EU without causing a rupture. On the other hand, Russia's growing political tensions with the EU at large are 'compensated' for by cooperation at other levels. Political tensions with the EU stimulate Moscow to develop a more extensive framework of bilateral relations, also in the field of energy, and to welcome certain regional programmes promoted by blocs of EU member states, with the Northern Dimension as perhaps the best example.[6]

In developing this argument, we build on the literature that stresses the multi-level nature of the EU–Russia relations where not only EU institutions but also member-state governments interact with the Russian government (Gower and Timmins, 2007; Smith, 2006; Maxine, Gower and Haukkala, 2013). We undertake a theoretical extension to this literature and argue that differences between the political systems in Russia and the EU lead to the distinctive abilities of multiple actors on each side to act in concert. Whereas institutionalized political processes on the EU side lead to largely predictable policy outcomes even when multiple actors are acting on behalf of the EU, this is not true of the Russian side. Russian political processes lack integration and coordination on virtually every dimension relevant to policy implementation. Weak political parties mean low accountability at election time and no policy mandate to be fulfilled afterwards. Sporadic and disjointed efforts at political mobilization mean that societal and group interests do not achieve a sustainable balance as they do in well-functioning democracies where the preferences of voter constituencies become aggregated through repeated political campaigns. Multiple levels of government and governments across the same level are not interlinked as regards the systematically accruing incentives to agents at those levels – not in any substantive partisan or electoral way. They are at best only loosely and inadvertently linked by the occasional administrative need or by personal contacts. As the policy objectives reflect conflicting interests of distinct and disjointed players, the policy making process in Russia has to deal with multiple concerns, and is often poorly balanced and unpredictable.[7]

---

6   EU initiative Northern Dimension put forward by Finland in 1997 and in 1999 approved by the EU Commission represents the most 'advanced' form of multi-level cooperation with participation of border regions. The ND could be seen primarily as an attempt on the part of interested EU members to avoid growing cleavage between the EU and Russia not by means of 'high politics' but by means of cooperation in solving practical functional problems.

7   Weeks (2012) argues that because of the inability to hold autocratic leaders accountable and the preferences of civilian elites personalist and military regimes should be the most conflict-prone; she finds that the only regimes that are as peaceful as democracies are non-personalist civilian regimes. See also Pickering and Kisangani, 2010.

To see whether Russian policies with regard to the post-Soviet space reflect contradictions between different objectives and considerations of Russian domestic actors, we focus on geopolitical considerations and the domestic political calculus. As the importance of various objectives and political interests behind them changes, we expect Russian policies to change as well. Such changes may be quite abortive at times, internationally damaging and may create long-term inefficiencies. As a rule, they are also inconsistent. In order to verify this assumption, we examine several abrupt Russian policy turns that were timed to the advantage of specific actors and their specific needs of the moment. We also hold that even the most puzzling policy turns were not detrimental to the interests of major Russian domestic players; further, that surprises tended to be smaller when the costs to those involved were higher, as was the case prior to the recent period of high prices for energy commodities.

Scholars and political observers have offered two types of explanations for recent tendencies toward conflict with the West over the post-Soviet space in Russian foreign policy. The first is that today's re-emergent Russia is striving to reassert itself as a global and regional power: as Russia grows stronger politically and economically, it becomes more active and confident in pursuing geopolitical goals (Aalto, 2008; Galbreath, 2008; Kanet, 2007; Mankoff, 2009; Wilson and Torjesen, 2008; Govella and Aggarwal, 2012; Sakwa, 2013). Viewing itself as the successor to 1000 years of Russian statehood (Morozov, 2008), Russia repeatedly endeavours to demonstrate its growing strength, seeking to convince the world that it is indeed a superpower. However, as Stent (2008: 1102) sums it up:

> an analysis of Putin's policy in the post-Soviet area raises questions about Russia's longer term interests. The Kremlin's maximum goal has been to secure recognition of this area as its legitimate sphere of influence that gives Russia a *droit de regard* over its former republics. To date, neither the West nor the CIS has obliged.

Even worse, it has been argued that attempts to become dominant regional power have led to contrary results and 'set in motion undesirable regional trends and exposed the weakness of its position in the multi-polar world' (Secrieru, 2009: 17). While the precise definition of Russian geopolitical goals differs, most authors agree that the pursuit of such goals leads to more tensions and conflicts with competing foreign powers (see Gomart, 2006; Legvold, 2007; Nygren, 2008; Póti, 2008; Tsygankov, 2008 and 2013).

The second approach explains Russia's foreign policy stance by the changes in Russia's domestic policies. The consensus among this group of scholars is that greater and more publicized assertiveness in foreign policy has been linked with the curbing of democracy in Russia since 2000 (Aslund and Kuchins, 2009; Blank, 2009; Busygina and Filippov, 2008; Hassner, 2008; Lukyanov, 2008; Okara, 2007; Filippov, 2009; Shevtsova, 2009). As Trenin (2009) argues, the uncompleted character of democratization in Russia is a major constraint for Western recognition of Russia's interests in the world, as such recognition is based on the quality of political, economic and social institutions and their democratic character. In the words of Dmitry Furman (2006: 73), 'the Cold War, which continues in disguised form, will stop only when Russia moves from managed democracy to democracy proper. If the structure of our society changes, then the entire system of our national interests will change as well'.

Theory of 'Diversionary Tension'

Students of international relations (IR) have often argued that state leaders may use foreign relations as an instrument for dealing with domestic political problems. The most popular and advanced version of this argument is the diversionary theory, which posits that incumbents may launch international conflicts in order to divert public attention from domestic issues. The theory underlying the general logic of diversionary conflict as applied in IR can be used for all types of political regimes, so this approach has also been employed to explain recent trends in Russian foreign policy.

We prefer a more refined version of diversionary conflict theory, aided by good insights offered by the literature. First, for diversionary conflict to be able to play any significant role, the leaders must be sensitive to shifts in public approval. Second, established democracies are less likely to start diversionary wars because, even if they might benefit from a diversion, the leaders would have difficulty in succeeding, due to the transparency of such societies. Although empirical studies of US behaviour during the Cold War show a connection between problems at home and aggressive behaviour abroad, for democracies in general such a relationship does not hold (Leeds and Davis, 1997). Third, as Goemans (2008) argues, diversionary logic works best for national incumbents who fear forcible removal from power. Similarly, Mansfield and Snyder (2002) have shown that the risk of military disputes is heightened when the countries involved are in the process of democratic transition: they are not fully democratic, but public opinion is not completely irrelevant. So while democracies have a lower risk of war, the risk for not-quite democracies is higher (Gleditsch and Ward, 2000). Fourth, as Morgan and Anderson (1999) have shown, diversion is effective only if the domestic audience exhibits some degree of internal cohesion; otherwise, international conflict might provoke rather than distract any domestic groups in active opposition to the incumbent administration. This may explain why, as the electoral strength of incumbents declines, they become more reluctant to create crises in foreign policy (Chiozza and Goemans, 2003; 2004). On the whole, 'states have tended to get into relatively more wars early in the election cycle and fewer wars late in the cycle' (Gaubatz, 1991: 212).

An important criticism of the diversionary argument is that it is hard to find suitable targets – enemy countries with low war costs and risk of failure (Levy, 1998). Critics have argued that the incumbents' political need for a diversionary war should be apparent also to potential targets, which would then take precautions (Clark, 2003; Fordham, 1998; Smith, 1996). However, environments of enduring rivalry make it possible to overcome this problem, and offer relatively more fertile sites for diversionary use (Mitchell and Prins, 2004). This suggests an important connection between geopolitical conflicts and the enhanced opportunity they create to use one's neighbours as the targets of diversionary conflicts.

Could Russian political incumbents be in need of a diversionary conflict or even a war? And might they be 'keeping' the post-Soviet space in a state of perpetual readiness for just such purposes? According to the literature, the conditions for such diversion are ripe and the regime is the most likely to use international conflicts for domestic ends when the regime is democratic but the quality of democracy is low, and when the regime is popular (especially early in the electoral cycle), and the government is uneasy about the threat of irregular removal from office. A suitable target of a diversionary conflict must be involved in long-term rivalry or, ideally, itself willing to engage in a conflict – possibly for its own diversionary reasons. So the answer to both questions is *yes:* Russian political incumbents may need a diversionary conflict; and they could be keeping the post-Soviet space in a state of perpetual readiness for that reason. The remainder of this chapter examines whether there is evidence to indicate that such diversionary strategies are in fact being used.

As Keohane and Milner (1996) explain, domestic politics cannot be understood without reference to the changing demands of the world economy and the links between the two. In all regimes, forces of international economics and globalization create opportunities, but they also significantly constrain social, political and economic actors. In non-democratic regimes, the trade-off is between the forced societal openness and greater pressure for political liberalization that such openness generates within the polity. As isolation of any kind will threaten to slow economic growth, posing a threat to regime stability in the long run, openness may seem indispensable, but it may also pose a challenge to regime control. The latter connection is so evident that, until recently, '… many Western policy makers and policy experts have assumed that political liberalization basically tracks the rate of economic growth, with only a slight lag, and that there is little that autocratic governments can do to stop it (as long as they remain committed to maintaining economic progress)' (Bueno de Mesquita and Downs, 2005: 77). However, those authors go on to note, that today's 'savvy autocrats, … have learned how to cut the cord between growth and freedom, enjoying the benefits of the former without the risks of the latter' (ibid). And as we argue, in the Russian case in particular, it is possible to track the exact mechanism (or at least one of the mechanisms) by which that can be done.

Wallander (2007) argues that the Kremlin is faced with both growing dependence on economic integration and increasing vulnerability to political influences from the outside, transmitted and mobilized in the modern information environment. Today it is not possible to put a lid on the flow of information or on the improvements and dissemination of information technologies, because that would stifle economic integration: the baby would get thrown out with the bathwater. 'The dilemma for the Kremlin is that the logic of its domestic political-economic system requires isolation, but sustaining power requires the wealth generated by participation in globalization, which would undermine that very system' (Wallander, 2007: 115). This view is fully consistent with the theory of disabling the mobilization of political opposition in autocratic regimes – openness makes it easier for opposition forces to identify themselves and organize for effective action, and even an apparently popular authoritarian government would want to ensure that such things do not happen. At the theoretical level, then, the dilemma noted by Wallander translates into strategic challenge for an authoritarian incumbent. As remarked by Bueno de Mesquita and Downs (2005: 86) '… to remain secure, autocrats must raise the costs of political coordination among the opposition without also raising the costs of economic coordination too dramatically since this could stymie economic growth and threaten the stability of the regime itself'.

To that argument we add analysis of a specific mechanism whereby the political regime in Russia has resolved such a dilemma, at least temporarily. Without resorting to informational isolation, it has succeeded in insulating domestic politics from 'harmful' international influences. This mechanism combines openness to the West with effective discrediting of all Western voices, by creating a virtual conflict with the West over a *third* area. With that done, the West is transformed into a biased actor in the eyes of the now-prejudiced domestic audience, its views and ideas compromised and rejected by the Russian public. At the same time, however, this discredited West remains, due to the very nature of its energy relationship with Russia, an important source of the revenues that secure the legitimacy and survival of the same incumbent regime that seems to have managed to 'inoculate' its own public against the possibly dangerous Western influence. This rejection of the West has been made possible by presenting the West as the main challenger to today's resurgent Russia in the post-Soviet space. What the official Russia presents as Western incursions into what is officially depicted as Russia's exclusive sphere of interest in the post-Soviet space has been used as an effective tool for rallying public support for the incumbent regime. Nowhere in the post-Soviet space is the rivalry between Russia and the West more visible than in Ukraine, which is seen by both Russian and Western 'geopoliticians' as a major geopolitical prize. This is also one reason why the Kremlin chose Ukraine as the area for employing its strategy of diversionary tension most effectively.

Why Pick on the Neighbours?

Looking for a safe instrument to sustain conflict with the West and thus block the dangerous political influence of Western criticism and democratic role-modelling, Russia's incumbent administration presented other post-Soviet countries as tangible agents of that remote arch-enemy. The Kremlin could not afford a real political conflict with the EU, the USA or any other major international player, no matter how helpful such a confrontation might have been for managing its domestic audience. It could afford occasional diplomatic tensions with a country like Great Britain – incidentally, at that time not a significant consumer of Russian energy. But to do that frequently would be risky, given the possibility of a coordinated response from the West. On the other hand, the Russian government had all reason to expect that carefully-engineered conflicts in the post-Soviet space would not cause any real breach with Western counterparts, who were seen as too dependent on Russian energy supplies and too cautious and lacking coordination in their policies towards Russia.[8] The resultant political tensions with the West would not be likely to reach a level harmful to trade relations. But to the domestic audience, and in light of Kosovo and Iraq, any consequent political rhetoric in the West against Russia's actions in its own neighbourhood could easily be presented as demonstrating not merely Western hypocrisy, but also Western expansionism – it could even be presented as a sign of imminent aggression or at least Western ideological incursion.

A conflict with a small post-Soviet state was a safe choice, but an opponent like that could not pose a serious threat to Russia. In fact, that was never what was claimed. Instead, the domestic audience was informed that those small countries were acting as proxies of more powerful and hostile international players. A manageable confrontation with a small country like Georgia could be real, with gunfire, and serve to draw in the public. The more useful confrontation with the West would remain 'virtual', in the media, but sufficient to serve the main goal of isolating Russia by making its population inattentive and hostile to outside voices. And what of Ukraine? That country is still viewed by many Russians as a cultural, social, historical and geopolitical 'extension' of their homeland. At the same time it has embarked, especially since the Orange Revolution, on a policy of seeking closer political, economic and institutional ties with the West. Thus Ukraine stood out as a preferred area for such a 'virtual' confrontation. As Ukraine is also an important area of transit of Russian energy to the West, it was only to be expected that sooner or later Russia would decide to play its 'energy card' in order to make this confrontation not only virtual but also real.

International Constraints

In terms of costs, any conflict with the post-Soviet countries, even the smallest among them, like the one with Georgia, comes with the risk of provoking not merely tensions but an actual rupture in relations with the West. Logically, the more intense the conflicts of the Russian government with post-Soviet countries, the more difficult will it be for the EU to disregard them. Such conflicts could also

---

8   Nevertheless, according to Feklyunina (2008: 619): 'the PR campaign has been trying to persuade the West that Russia does not cherish any imperial ambitions regarding the CIS countries'.

be expected to have direct impact on the EU's economic and political relations with Russia. In general, the reaction of the EU to Russian actions could be influenced by the multi-level interactions that the EU – and its individual member countries – have with Russia (Smith, 2006). Given such a strategic linkage, active programmes existing at one level could explain decisions made at other levels. It is also reasonable to assume that while some actors in Europe focus more on relations between Russia and the EU as whole, other key countries may be more interested in good bilateral relations or economic cooperation. Similarly, the various EU players may focus on different issues – like security, environmental issues, energy supplies or democratic development. In general we could expect that Russia's EU counterparts would define their strategies, by, first, taking into account what happens at all levels of interactions with Russia, and second, by assigning different priorities to interactions at different levels and in different areas. This means that the Council of the EU, the European Commission, the European Parliament, individual member states, regional blocs and sub-national governments of the member states are all likely to pursue different and even contradicting strategies vis-à-vis Russia. Moreover, the same actor may well adopt different strategies for different institutional playing fields.

Above we argued that the Russian leadership has a motive for sustaining conflicts with many post-Soviet countries even at the price of increasing political tensions with the West – because such conflict can provide specific benefits of 'political isolation'. However, regardless of exactly why the tensions between the West and Russia arise, the connectedness among the many lines of Russia's interaction with the outside world can result in a counter-intuitive trade-off in Russian foreign relations: better economic relations with the West could, paradoxically enough, lead to more tension in political relations. For instance, it seems that the high degree of mutual energy-trade dependence between EU and Russia makes the Russian government able to afford more tension without it coming to actual rupture. In addition, the growing political tensions with the EU at large could be 'compensated' for by collaboration at other institutional levels, or by stepping up bilateral intergovernmental cooperation. Political tensions with the EU could thus stimulate the Russian government to develop a more extensive framework of bilateral relations and to welcome regional programmes promoted by blocs of EU member states. Importantly, however, it also follows that, for Russia to be able to play this 'counterbalancing' role, the content of bilateral relations and regional programmes must be strictly isolated from those issues which serve as the pretext for EU–Russia tensions. The bilateral relations are less suited to be used in dealing with politically charged issues like the development of democracy in Russia, restrictions on political competition, freedom of the press or human rights (Busygina and Filippov, 2009).

For example, the French President is expected to act differently in the European Council, where he acts in a multilateral framework, and in a bilateral Russian–French context. Similarly, optimal strategies for Finland are likely to differ, depending on whether the setting is the European Council, the Northern Dimension framework, when Finland presides over the OSCE, or in bilateral dealings with Russia. This apparent lack of coherence of EU foreign policy is a natural reflection of its multilevel governance structure, and not so much a product of Russia trying to undermine unity though bilateral dealings.

The Russian government can be particularly successful in applying differentiated approaches to bilateral and multilateral relations with Western countries because the Kremlin enjoys freedom from ideological constrains, 'norms', 'values' or, if successful in its diversionary tactics, from the pressure of public opinion. Today the Kremlin can implement diametrically opposed policies in different venues in its European relations, without repercussions from its domestic audience. Illustrative here is the contrast between strained Estonian–Russian relations and the friendly tone in relations between Finland and Russia.

Through the twentieth century, Russian–Finnish relations were ridden with mutual territorial claims, interventions and population repatriation, not mention the Winter War and the Continuation War and Finland fighting in the Second World War on the Axis side against the Soviet Union. However, simple comparison of how the Russian media portray relations with Finland and with Estonia reveals a striking contrast. Finland is always depicted in a highly positive, friendly light. Russian television regularly reminds viewers about the crimes of Estonian Nazi collaborators, but it never portrays the Finns that way. And no mention is made of the generally poor image of Russia in modern Finland (Gallup polls in 2004 showed that 62 per cent of Finns had a negative opinion of Russia). This shows impressive restraint on the part of the journalists, since worse attitudes towards Russia are reported only in Kosovo. Surprisingly, Kosovo and Finland are almost never linked in the Russian media when it criticizes the role of Martti Ahtisaari in the Kosovo conflict.[9] Apparently, as Timo Vihavainen has pointed out, 'history plays a very different role in Finnish–Russian relations than it does in Polish–Russian or Estonian–Russian relations' (Vihavainen, 2006: 28). When visiting Finland in 1997, President Yeltsin apologized for 'the Soviet aggression'. President Putin went even further and laid a wreath on the grave of Marshal Mannerheim during his visit to Finland in 2001. Though it caused strong reactions in Finland, the 'three R-words' statement of Finnish Defence Minister Jyri Häkämies in 2007 in Washington – when he said that 'the three main security challenges for Finland today are Russia, Russia and Russia; and not only for Finland, but for all of us' – went almost unnoticed in the Russian media. Nor was any attention drawn to the fact that Finland, normally a strong supporter of international initiatives, declined to sign the Ottawa Treaty banning anti-personnel landmines, which are used on the Finnish–Russian border.

---

9   In 1999, Martti Ahtisaari (President of Finland, 1994–2000) crafted proposals for resolving the conflict in Kosovo. From autumn 2005, Ahtisaari headed talks on the future status of Kosovo. Without doubt, he has received more publicity in Russia than any other Finn in recent memory. For an example of the Russian media reaction toward Ahtisaari, see an interview by Dmitry Rogozin for the programme Vesti on 10 October 2008, discussing the 'creativity of Mr Ahtisaari' that 'broke international law'. Available at: http://www.vesti.ru/doc.html?id=215270.

Moreover, although Finland was serving as OSCE chair during the war in Georgia and therefore had to react critically, there were practically no negative attacks against 'our northern neighbour' in the Russian media. However, this positive attitude toward Finland has changed lately: during the autumn 2012 Russian media had a campaign of several weeks aimed against Finland, claiming that Finnish social welfare authorities are discriminating and treating in an inhumane way children of Russian descent. By the end of 2012, child-related issues had become the high-interest media topic in Russia, with the focus on human rights abuses of Russian children in the USA, especially after Russian banned adoption by US citizens.

Russia is essentially still at the stage of setting its basic constitutional principles, the choice being between a competitive democratic system and an authoritarian model. The period of high prices of energy and natural resources in the early 2000s made the authoritarian model attractive for the Kremlin, consistently with the argument that energy prices are the driving factor behind the directional swings in Russia's domestic development (Gaidar, 2007). It seems that for Russia, high energy prices and high revenues from energy exports create wrong incentives, not only for the country's own economic and political development (the 'resource curse') but for its foreign policy as well. Indeed, with the high energy prices after the turn of the millennium, EU–Russia relations began to exhibit the growing disconnect between their economic and political components: as EU reliance on trade with Russia increased, political relations deteriorated. Moreover, energy security began to define the most important lines of tension. Especially Russia's stoppage of gas supplies for Ukraine in 2006, in the wake of the Orange Revolution during which Russia and President Putin personally were humiliated, caused new tensions and led many Western policy makers to fear that energy dependence on Russia could pose a challenge to the security of their countries. However, Europe had no other options but to continue to buy Russian gas and oil, which in turn made Russia not only richer in economic terms, but also more challenging in political terms. The second round of the gas conflict between Russia and Ukraine in 2009 transformed this 'virtual diversionary tension' into an open conflict of economic, political and security interests.

The decline in energy prices caused by the global economic crisis might be expected to lead to the 'normalization' of Russia, with it becoming more willing to accommodate the interests of others. However, our argument here points to a different possibility. Since the task of 'diversion' remains a driving force in Russia's foreign relations and, if anything will rise in importance as the population experience growing economic hardships, the decline in trade benefits might turn the Kremlin to even more risky policies, in attempts to insulate Russia even further against international influences. The fact that Russia was facing a new electoral cycle with many uncertainties in 2011 and 2012 could make it more willing to take greater risks. The outburst of the anti-American and anti-Western rhetoric during Putin's electoral campaign was indeed a clear illustration of that negative trend. At the same time, however, some among the country's political elite had realized that Russia needs more cooperation and not more tension with the West, and that resulted in the adoption of a more conciliatory and less confrontational approach and the signing of several cooperation deals between Russian and Western energy companies. On the other hand, the situation in Ukraine may have developed in such a direction that tensions between Russia and Ukraine are no longer a function of Russia merely playing an anti-Western card for domestic purposes, but an expression of a deeper conflict of interests between Moscow and Kiev.

Theory, however, suggests that there is a binding constraint on the potential 'diversionary belligerence'. The diversionary argument as presented in the IR literature (Smith, 1996; Fordham, 1998) applies to democracies, or at least to the types of political arrangements where public support is essential to government survival. Thus, as Russia moves closer to autocracy, it should be less likely to 'divert' and to use foreign policy for domestic purposes. Russia today can be described as a semi-democratic political system combined with open market economy. If it moves towards a more authoritarian system, the utility of 'virtual' conflict with the West will decline, and Russian foreign policy will come to resemble a more pragmatic version of the 'Cold War' period.

What then might Russia's Western partners do in the short term? If Russia is looking for a confrontation for its own purposes, it would be wise for the West to oblige, while maintaining some level of critical rhetoric against it – otherwise, the Kremlin might be forced to engineer new crises. But here we may expect some division of labour between the USA and Europe. One angry voice from the West serves the purpose just as well as two, so the European Union, with its numerous practical cooperative projects with Russia to protect, may opt to stay less involved, trusting the US side to do most of the talking.

Continued Tension in Russia–EU Political Relations

For many reasons, relations between Russia and the EU as a whole are likely to remain tense. Politicians on either side can expect few benefits from seeking compromises. On the contrary, in both Russia and the EU, politicians can rely on the continuing tension as a mechanism for generating internal consensus, while they go about implementing the transformations of their respective political systems.

The political systems in the EU and Russia as well are experiencing a period of transition, albeit with very different starting points. The high price of energy resources on the international market has made the authoritarian model attractive and viable for the Kremlin, making it less cooperative or willing to accommodate the interests of others. Unfortunately, the recent global financial crises failed to create sufficient incentives for change – whether in the structure of the economy, or in political institutions. With lower energy prices came debate on the need for modernization and diversification of the Russian economy, making the leadership realize how vulnerable the country was, even spurring some political figures to call for political liberalization and real democratization as a remedy. However, with oil prices going up in the wake of the 'Arab Spring' revolutions and global economic recovery, and with a new State Duma and presidential elections, this talk of modernization, diversification and liberalization did not become a walk towards modernization, diversification and liberalization. It seems that much is set to remain the same in Russia, and that Russia's relations with the West will follow a bumpy road.

Energy trade with the EU is crucial to the Russian economy. Economic objectives alone would require advancing cooperative relations with EU institutions as well as with most EU member states. Unfortunately, a combination of geopolitical goals and domestic political considerations has motivated the government to continue picking numerous fights with the EU, the USA and most post-Soviet states. These conflicts help the Kremlin to legitimize its authoritarian inclinations vis-à-vis the domestic public, while continued cooperation allows it to profit from selling natural resources to the West at high prices.

In short, in the case of Russia, domestic political calculations often encourage key players to wage uncompromising rhetorical or 'virtual' battles with various foreign counterparts, aiming primarily to reach the domestic audience. The battles in these virtual conflicts are played out chiefly on government-controlled television channels, with politicians and pundits competing to present the most vivid and creative extremist stances. They convey an image of 'the Great Russia' as surrounded by enemies and suffering from the biases of Western media (Godzimirski, 2009; Mankoff, 2012; Ziegler, 2012).

EU's Strategic Reaction

Unfortunately for EU–Russian relations, Moscow's domestic political rhetoric and its hardliner attitude towards the former Soviet republics make Russia a prize candidate to fulfil the emerging political demand in Europe for a unifying external threat. Russia is too large and unpredictable a country to be ignored, and its role as an important energy supplier automatically puts it on the strategic maps of most member states and of the EU as a whole. Despite impressive success in economic integration, the very principles, forms and limits of political integration in the EU are still subject to debate among and within its members. The design and implementation of a successful common foreign policy could mark a crucial step in the EU's constitutional development, as it could be used to justify further expansion of the political powers of EU-level institutions. However, foreign policy is also an area where member states tend to disagree. Anticipating such disagreements over policies, member states are also likely to disagree on whether to delegate foreign policy decisions to the EU level. Constitutional steps toward the development of a joint foreign policy require there to be a crucial issue on which there is unwavering consensus among member states – as Romano Prodi aptly noted, 'Europe needs a sense of meaning and purpose'.[10]

---

10   R. Prodi. Shaping the New Europe. Address to the European Parliament, 15 February. Strasbourg, 2000.

Conflict with Russia – or at least the perception of Russia as a challenging partner that threatens the energy and security interests of many countries – is an attractive candidate for such a unifying issue, a basis that could produce the consensus needed to give the EU stronger powers in foreign policy (DeBardeleben, 2012; Klinke, 2012). In principle, the necessary consensus would have been equally feasible for the issue of fostering democracy in Russia, and for a while this seemed to be the case. But by 2000, the situation in Russia had changed drastically, and as the democratization trend was reversed, the grounds for such consensus vanished. After that, the strategy of confronting Russia became the only consensus-based option for the EU institutions, particularly for the Commission and the European Parliament. As energy cooperation with Russia is the backbone of that relationship and is seen as posing a strategic challenge to the EU – and to several EU member states – it is hardly surprising that the EU should take energy measures to cope with this new emerging common task. As shown by the adoption of the 'Third Package' and the renewed EU focus on energy infrastructure, the European Union has indeed begun adapting to this new strategic energy environment – and here Russia remains an important element.

Conclusions

Until at least very recently, EU–Russian relations were developing in a paradoxical way: as the dependence of both sides on mutual trade increased, their political relations deteriorated. This chapter has aimed at explaining why cooperation between Russia and the EU has remained so severely limited in many areas of mutual interest.

The deterioration of the EU–Russian relations is, in our opinion, a result of strategic choice. We connect the dynamics of the EU–Russia relations with the effect of high energy prices and the growing importance of energy trade (especially in natural gas) between Europe and Russia. The high share in Russia's national revenues from exports of natural resources seems to have created specific type of incentives for economic development, growth, political system and – not least – foreign policy. The Russian leadership has domestic political incentives for sustaining a certain level of political conflict with the EU and many post-Soviet countries. These incentives stem precisely from the importance of continued cooperation and openness on other issues. This 'virtual conflict', in turn, provides an opportunity to isolate Russia's domestic politics from international influences without having to resort to economic and informational isolation. It helps to combine both openness to the West (important for maintaining an open market economy), and the effective silencing of any Western critics.

References

Aalto, P. 2008. *The EU–Russian Energy Dialogue: Europe's Future Energy Security*. Aldershot: Ashgate.

Arezki, R. and M. Bruckner 2011. Oil rents, corruption, and state stability: evidence from panel data regressions. *European Economic Review*, 55(7), 955–63.

Aslund, A. and A. Kuchins 2009. *The Russia Balance Sheet*. Washington, DC: Peterson Institute for International Economics, Center for Strategic and International Studies.

Balmaceda, M.M. 2008. *Energy Dependency, Politics and Corruption in the Former Soviet Union: Russia's Power, Oligarchs' Profits and Ukraine's Missing Energy Policy, 1995–2006*. London: Routledge.

Beblawi, H. and G. Luciani 1987. *The Rentier State*. New York: Croom Helm.

Blank, S. 2009. *What Is Living and What Is Dead in the Russian Foreign Policy Tradition*, Paper presented at the Annual Convention of the International Studies Association, New York, 15–18 February 2009.

Boussena, S. and C. Locatelli 2013. Energy institutional and organisational changes in EU and Russia: revisiting gas relations. *Energy Policy*, 55(April), 180–89.

Bueno de Mesquita, B. and G.W. Downs 2005. Development and democracy. *Foreign Affairs*, 84(5), 77–86.

Busygina, I. 2007. Russia's regions in shaping national foreign policy, in *Russia and Europe in the Twenty-First Century: An Uneasy Partnership*, edited by J. Gower and G. Timmins. London: Anthem Press, 75–88.

Busygina, I. and M. Filippov 2008. End comment: EU–Russian relations and the limits of the Northern Dimension, in *The New Northern Dimension of the European Neighborhood*, edited by P. Aalto, H. Blakkisrud and H. Smith. Brussels: Centre for European Policy Studies, 204–19.

Chiozza, G. and H. Goemans 2003. Peace through insecurity: tenure and international conflict. *Journal of Conflict Resolution*, 47(4), 443–67.

Chiozza, G. and H. Goemans 2004. Avoiding diversionary targets. *Journal of Peace Research*, 41(4), 423–43.

Clark, D.H. 2003. Can strategic interaction divert diversionary behavior? A model of U.S. conflict propensity. *Journal of Politics*, 65(4), 1013–39.

DeBardeleben, J. 2012. Applying constructivism to understanding EU–Russian relations. *International Politics*, 49(4), 418–33.

Feklyunina, V. 2008. Battle for perceptions: projecting Russia in the west. *Europe-Asia Studies*, 60(4), 605–29.

Filippov, M. 2009. Diversionary role of the Georgia–Russia conflict: international constraints and domestic appeal. *Europe–Asia Studies*, 61(10), 1825–47.

Fordham, B. 1998. Partisanship, macroeconomic policy, and US uses of force, 1949–1994. *Journal of Conflict Resolution*, 42(4), 418–39.

Friedman, T. 2006. The first law of petropolitics. *Foreign Policy*, 154, 28–36.

Furman, D. 2006. A silent Cold War. *Russia in Global Affairs*, 4(2), 68–74.

Gaidar, E. 2007. *Collapse of an Empire: Lessons for Modern Russia*. Washington, DC: Brookings Institution Press.

Galbreath, D.J. 2008. Putin's Russia and the 'new Cold War': interpreting myth and reality. *Europe–Asia Studies*, 60(9), 1623–30.

Gaubatz, K.T. 1991. Election cycles and war. *Journal of Conflict Resolution*, 35(2), 212–44.

Gleditsch, K.S. and M.D. Ward 2000. War and peace in space and time: the role of democratization. *International Studies Quarterly*, 44(1), 1–29.

Godzimirski, J. 2009. Energy security and the politics of identity, in *Political Economy of Energy in Europe: Forces of Integration and Fragmentation*, edited by G. Fermann. Berlin: Berliner Wissenschaftsverlag, 173–206.

Goemans, H. 2008. Which way out? The manner and consequences of losing office. *Journal of Conflict Resolution*, 52(6), 771–94.

Gomart, T. 2006. *Russian Foreign Policy: Strange Inconsistency*. Swindon: Conflict Studies Research Centre, Defence Academy of the United Kingdom.

Govella, K. and V.K. Aggarwal 2012. *Responding to a Resurgent Russia: Russian Policy and Responses from the European Union and the United States*. SpringerLink Bücher.

Haber, S. and V. Menaldo 2011. Do natural resources fuel authoritarianism? A reappraisal of the resource curse. *American Political Science Review*, 105(1), 1–26.

Hassner, P. 2008. Russia's transition to autocracy, *Journal of Democracy*, 19(2), 5–15.

Kanet, R. (ed.) 2007. *Russia: Re-emerging Great Power*. New York: Palgrave Macmillan.

Karl, T. 1997. *The Paradox of Plenty: Oil Booms and Petro-States*. Berkeley: University of California Press.

Keohane, R. and H. Milner 1996. *Internationalization and Domestic Politics*. Cambridge: Cambridge University Press.

Klinke, I. 2012. Postmodern geopolitics? The European Union eyes Russia. *Europe–Asia Studies*, 64(5), 929–47.

Krastev, I. 2008. *The Crisis of the Post-Cold War European Order: What to Do About Russia's Newfound Taste for Confrontation with the West*. Washington, DC: The German Marshall Fund of the United States.

Leeds, B. and D. Davis 1997. Domestic political vulnerability and international disputes. *Journal of Conflict Resolution*, 41(6), 814–34.

Legvold, R. (ed.) 2007. *Russian Foreign Policy in the Twenty-First Century and the Shadow of the Past*. New York: Columbia University Press.

Levy, J. 1998. The causes of war and the conditions of peace. *Annual Review of Political Science*, 1(1), 139–65.

Lukyanov, F. 2008. Interactions between Russian foreign and domestic politics. *Irish Studies in International Affairs*, 19(1), 17–24.

Mankoff, J. 2012 *Russian Foreign Policy: The Return of Great Power Politics*. Lanham, MD: Rowman and Littlefield.

Mansfield, E.D. and J. Snyder 2002. Incomplete democratization and the outbreak of military disputes. *International Studies Quarterly*, 46(4), 529–49.

Maxine D., J. Gower and H. Haukkala (eds) 2013. *National Perspectives on Russia: European Foreign Policy in the Making?* New York: Routledge.

Mitchell, S. and B. Prins 2004. Rivalry and diversionary uses of force. *Journal of Conflict Resolution*, 48(6), 937–61.

Morgan, T.C. and C. Anderson 1999. Domestic support and diversionary external conflict in Great Britain, 1950–1992. *Journal of Politics*, 61(3), 799–814.

Morozov, V. 2008. Europe: orientation in time and space. *Russia in Global Politics*, 3.

Nygren, B. 2008. *The Rebuilding of Greater Russia: Putin's Foreign Policy Towards the CIS Countries*. New York: Routledge.

Okara, A. 2007. Sovereign democracy: a new Russian idea or a PR project? *Russia in Global Affairs*, 5(3), 8–20.

Pickering, J. and E.F. Kisangani 2010. Diversionary despots? Comparing autocracies' propensities to use and to benefit from military force. *American Journal of Political Science*, 54(2), 477–93.

Póti, L. 2008. Evolving Russian foreign and security policy: interpreting the Putin Doctrine. *Acta Slavica Iaponica*, 25, 29–42.

Roberts, C.A. 2007. *Russia and the European Union: The Sources and Limits of 'Special Relationships'*. Carlisle Barracks, PA: Army War College Strategic Studies Institute.

Ross, M.L. 2012. *The Oil Curse: How Petroleum Wealth Shapes the Development of Nations*. Princeton, NJ: Princeton University Press.

Secrieru, S. 2009. Illusion of power: Russia after the south Caucasus battle, CEPS Working Document No. 311. Available at http://www.ceps.eu/book/illusion-power-russia-after-south-caucasus-battle [accessed: 15 June 2012].

Sakwa, R. 2013. The cold peace: Russo-Western relations as a mimetic cold war. *Cambridge Review of International Affairs*, (ahead-of-print): 1–22.

Semenenko, I. 2013. The quest for identity. Russian public opinion on Europe and the European Union and the national identity agenda. *Perspectives on European Politics and Society*, (ahead-of-print): 1–21.

Shevtsova, L. 2009. The return of personalized power. *Journal of Democracy*, 20(2), 61–5.

Smith, A. 1996. Diversionary foreign policy in democratic systems. *International Studies Quarterly*, 40(1), 133–53.

Smith, H. (ed.) 2006. *The Two-Level Game: Russia's Relations with Great Britain, Finland and the European Union*. Helsinki: Aleksanteri Institute.

Smith, M.E. and M. Webber 2008. Political dialogue and security in the European neighbourhood: the virtues and limits of 'new partnership perspectives'. *European Foreign Affairs Review*, 13(2), 73–95.

Stent, A.E. 2008. Restoration and revolution in Putin's foreign policy. *Europe-Asia Studies*, 60(6), 1089–106.

Stratfor, 2013. *The Past, Present and Future of Russian Energy Strategy*. Available at: http://www.stratfor.com/weekly/past-present-and-future-russian-energy-strategy [accessed: 12 March 2013].

Treisman, D. 2011. *The Return: Russia's Journey from Gorbachev to Medvedev*. New York: Simon & Schuster.

Trenin, D. 2009. *Odinochnoe Plavanie [Lonesome Swimming]*. Moscow: Moscow Carnegie Endowment.

Tsygankov, A. 2008. Russia's international assertiveness: what does it mean for the West? *Problems of Post-Communism*, 55(2), 38–55.

Tsygankov, A. 2013. The Russia–NATO mistrust: ethnophobia and the double expansion to contain 'the Russian bear'. *Communist and Post-Communist Studies*, 46(1), 179–88.

Victor, D.G., A. Jaffe and M.H. Hayes. 2006. *Natural Gas and Geopolitics: from 1970 to 2040*. Cambridge: Cambridge University Press.

Wallander, C. 2007. Russian transimperialism and its implications. *The Washington Quarterly*, 30(2), 107–22.

Wantchekon, L. 2002. Why do resource dependent countries have authoritarian governments? *Journal of African Finance and Economic Development*, 5(2), 57–77.

Weeks, Jessica 2012. Strongmen and straw men: authoritarian regimes and the initiation of international conflict. *American Political Science Review*, 106(2), 326–46.

Wilson, E. and S. Torjesen 2008. *The Multilateral Dimension in Russian Foreign Policy*. London: Routledge.

Ziegler, C. 2012. Conceptualizing sovereignty in Russian foreign policy: realist and constructivist perspectives. *International Politics*, 49(4), 400–417.

Chapter 6

Diversification, Russian-style: Searching for Security of Demand and Transit

Pavel K. Baev

Introduction

From the early 2000s, the proposition that the demand for hydrocarbons, as well as their price, would rise indefinitely had acquired in Russia the convincing power of not only a winning business strategy but also a dominant political philosophy. Then came the shocking collapse of prices and demand in the second half of 2008 that undermined this philosophy of 'energy superpower' – but adjustments in energy policy have been marginal and reluctant. This process of internalizing undesirable changes has been impeded by unwillingness to contemplate what a possible end to petro-prosperity might mean for the durability of the existing political system, based as it has been on an unwritten social contract for ever-increasing rewards, and supported by climbing oil prices (Makarkin, 2011). The gradual recovery of oil prices to the plateau of USD 100–120 per barrel by mid-2011 (Brent crude benchmark) could not compensate for the populist over-spending of the 2011/2012 election cycle, leaving Russia highly vulnerable to new shocks (Aleksashenko, 2012; Guriev and Tsyvinski, 2010). Energy policy, in the meantime, has remained focused on mega-projects aimed at securing significant increases in the production and export of oil and natural gas. The goals of that policy were outlined in the 2009 *Energy Strategy until 2030* (Russian Government, 2009) but that document was fast overtaken by events, and no new guidelines have been issued; even Putin had very little to say about the energy prospects in his 'visionary' articles published in the course of the election campaign (Putin, 2012).

With the start of Putin's new presidential term, Russia has entered into a profound political crisis that deforms its system of governance and affects every aspect of decision making, including the dirigisme that has shaped the development of the energy sector. The interplay between political turmoil and energy issues has become puzzlingly complex indeed. The various contributors to this volume seek to isolate particular angles of this phenomenon, generally agreeing on the crucial importance of the performance of the oil/gas sector for Russia's ability to gain new dynamism and achieve a breakthrough in modernization. The analysis in this chapter focuses on the increasingly acute need in changing the pattern of energy export, and argues that the consolidation of this pattern was determined by the establishment of the state-centric economic model of Putinism. The massive outflow of capital in 2012 and the economic slowdown in early 2013 show that the petro-rent distribution paradigm has exhausted its growth potential, and a reconfiguration of energy flows is necessary for transforming this outdated model.

One specific impact of the fusion of political and economic turmoil has been greater separation of elements of energy sector, in particular the divergence of the oil and gas markets. This means better prospects of de-linking the slowly expanding gas market, where the spot business is gaining ground, from the uncontrollable and unpredictable oil price (Stern, 2010). Gazprom has stubbornly resisted such an unstoppable liberalization of the gas market, landing the Russian 'champion' in serious trouble with the European Commission (Kulikov, 2012a). The gas sector in Russia is far more politicized and accustomed to enjoying greater personal attention from Vladimir Putin than the oil business, and the unexpected erosion of Gazprom's position in the European gas market in 2012 brought sharp political remonstrations. Nevertheless, both the economics of production and clan intrigues continue to tie the oil and gas businesses together, so it makes sense in the further analysis in this chapter to examine the role of these two main hydrocarbon commodities together.

Another heavy impact of the global crisis is the accelerated divergence of the major energy markets, where demand has been fluctuating in remarkably dissimilar patterns. The US economy was initially the main generator of the troubles, and the recovery has been uncertain, but the contraction happened to coincide with the phenomenon known as the 'shale gas revolution'. Since 2010, the USA has become the top producer of gas, primarily due to its fast-expanding exploitation of domestic deposits of non-conventional gas. Imports have shrunk to negligible volumes, and the country might soon become a net exporter (IEA, 2009: 400–415; IEA, 2011: 169). Europe has been affected severely but unevenly: the German economy remains strong, but the deep recession in the 'southern tier' sets a hard test for the EU institutions and derails the process of European integration. China weathered the first crisis typhoon easily, and has shown signs of a slowdown only in 2012; but its demand for primary energy resources has been steadily expanding. Russia, despite its accumulated financial reserves, has been among those worst affected by the crisis. The key reference point here is neither the G7 nor the BRICS economies, but the main petro-producers, like Saudi Arabia, which experienced a painful shock in autumn 2008 but have rebounded swiftly. The official explanation of this L-shaped recession is far from convincing, but one lesson learned has definitely been the imperative of greater flexibility in energy export flows (see Mitrova's chapter in this volume; also Baev, 2010b).

This chapter examines the evolving policy of diversifying Russia's energy export shaped by the interplay between political preferences, court intrigues and economic rationale (or the lack of thereof). It starts with an overview of fluid guidelines for this policy; then looks into the prospects in two main parts of the European market: Germany (focusing on the role of the undersea Nord Stream pipeline), and Southeastern Europe plus Turkey and Ukraine. Next, the shift towards the Asia–Pacific markets is evaluated, with particular attention to China. The conclusion offers some informed guesses about a possible re-orientation of Russian energy destinations.

The Meaning of the D-word

Diversification of supply has become a key element of EU energy strategy (to the extent that such a correspondence of ends and means exists) since the first Russian–Ukrainian gas skirmish in January 2006, but Russia has entertained its own ideas in this regard for at least as long. In September 2006, President Vladimir Putin impressed participants at the Valdai Club meeting when he informed them that Russia would direct to Asia 'around 30 per cent of our total energy export over the next 10–15 years', as against a mere 3 per cent at that time (Putin, 2006). That goal was clearly unattainable and is indeed nowhere close to being achieved seven years later, but its declaration was more than just a bluff intended to mislead excitable Europeans.

The material substance of the 'go-East' vision includes the plans for developing Russia's vast but poorly explored oil and gas reserves in East Siberia and the Far East, which in practice could be exported only towards the expanding Asia–Pacific markets. Realization of these plans, however, requires massive investments and many years of implementation, stretching beyond the foreseeable political horizon (Milov, 2006). Indeed, as of early 2013, only one group of projects had been successfully implemented in that enormous region – on Sakhalin island – and the credit for bringing these challenging projects to fruition goes to consortiums of Western oil 'majors' that designed them on the basis of production-sharing agreements (PSAs) from the mid-1990s (Krysek, 2007). Realistically, only one big project could have been executed in the early 2010s – the Kovykta gasfield to the north of Lake Baikal – but that venture has fallen prey to Gazprom's desire for market dominance (Bradshaw, 2008; Naumov, 2011).

Procrastination in establishing a new production base to the east of the Yenisei River has meant that there are only two ways in which Russia could boost its energy export to Asia–Pacific: either commercially – by buying all available oil and gas in Central Asia for re-export; or politically – by diverting some of the energy flows from the European market. The choice is directly connected with another dilemma: whether Russia should invest in the development of own oil and gas fields in the Far East and the Far North, or put money into joint ventures in the Caspian area. The second option had seemed more cost-efficient, less technologically challenging and highly advantageous in geopolitical terms, so throughout much of the first decade of the 2000s, the Kremlin gave it prime consideration (Overland, Kjaernet and Kendall-Taylor, 2009; Baev, 2008). In 2005–2007, Putin put great personal efforts into negotiating a series of deals with counterparts from Kazakhstan, Turkmenistan and Uzbekistan on purchasing all available volumes of gas and constructing a new pipeline along the eastern shore of the Caspian Sea.

What brought that gas offensive to an abrupt end was the sharp downturn in Russia's economic fortunes in the second half of 2008, when Gazprom suddenly discovered that holding its part of the 'take-or-pay' deals was proving prohibitively expensive because the arrangement on 'European' prices could not be undone. The pipeline explosion in Turkmenistan in January 2009 was used as a pretext for stopping all import of gas from this major producer. When a new deal was struck a year later, the contracted volume was reduced to just 11 bcm (Smirnov, 2010). Agreements with Kazakhstan and Uzbekistan, with the generous price formula, have been honoured, but a significant reduction of Russia's energy interests in Central Asia has been registered by all actors in this far-from-great game (Satayev, 2010). Perhaps the most evident manifestation of this down scaling concerns the gas pipeline from the Saman-Depe field on the right bank of the Amu-Darya in Turkmenistan to Alashankou in Xinjiang, China, which by the end of 2011 had reached planned capacity of 40 bcm a year, carrying exactly the volumes previously contracted by Russia (Aliev, 2009; Wong, 2011; O'Hara and Lai, 2011).

This retreat from the South – partial and conditional – does not signify any curtailing of the proposition concerning greater diversification of energy export: rather, it confirms its hasty revision. 'Oil goes East, gas goes West' appears to be the bottom line of today's thinking. Thus, construction of the East Siberia–Pacific Ocean (VSTO–ESPO) oil pipeline was realized at a Stakhanovite tempo, whereas the Altai gas pipeline has remained in the 'interesting-idea' phase (Ponomarev, 2010a; Zhiznin, 2011).

Almighty Gazprom had to extend its timetable for developing the Bovanenkovo gasfield on the Yamal, but still succeeded in getting this top priority project on-line, while working feverishly to recapture its lost positions on the European market. A major goal here has been to reduce transit dependency by building new pipeline routes with spare capacity, but this plan has clashed with the aim of expanding economic cooperation with Belarus and Ukraine, besides being hugely expensive. Developing capacity for exporting liquefied natural gas (LNG) was another major goal in Gazprom's diversification package, but the main delivery target was to be the US market, which to all intents and purposes has disappeared, so currently it is NOVATEK that is advancing a large-scale LNG project, while Gazprom has lost interest in this politically-indifferent business avenue (Kupchinsky, 2009; Mitrova, 2011).

Overall, the strengthening of government control over Russia's energy sector, including the oil industry, from 2006 onwards provided the conditions for developing a long-term strategy for enhancing energy security by diversifying export flows. However, nothing resembling such a strategy has actually appeared, and the current course seems to be going stubbornly against the tide of developing trends on the world gas market (as discussed, for instance, in Lunden and Fjærtoft's chapter in this volume).

Keeping Germany Hooked, the EU Confused, and Belarus Tamed

Of all the energy markets it is Germany that has been pivotal for Russia, ever since Soviet times: Not only because of the volumes of gas headed there, but also because this deep interdependency reproduces the model of long-term contracts based on price formulas for gas tied to fluctuations in the oil price (Stern, 1999; Khodov, 2010).

This well-managed gas trade has helped Gazprom to cultivate partnerships with two major German corporations – E.ON-Ruhrgas and BASF-Wintershall; it has also created an opportunity for Putin to build special relations with the German political elite (Rahr, 2007). The last thing that the Russian leadership could possibly aim at is diversifying gas export away from Germany – but this calculated commitment is not necessarily shared by the ruling coalition in Berlin. Many German politicians, including Chancellor Angela Merkel, are displeased and even embarrassed by Germany's role as Russia's 'best friend', and influential business interests are frustrated with the ever-deepening corruption in Russia (Stent, 2007). Unmistakable evidence of this frustration came with the decision of E.ON to sell its 3.5 per cent shares in Gazprom, which meant that its representative would no longer sit on the board of the Russian super giant (Grib, Egikyan and Kiseleva, 2010).

While Russia firmly rejected the Energy Charter as the legal foundation of gas trade with the EU, it was also worried about the stalemate on the energy 'road' of the 'strategic dialogue' with the Union, and was quite unhappy about the legislation and decision making in Brussels that took no heed of Russian preferences. Putin therefore counted on German support, as well as on backing from France and Italy, in sabotaging and remaking the EU energy policy – which in the Russian reading makes no economic sense whatsoever, besides being explicitly Gazprom-phobic. The famous European Renewable Energy Directive (the '20–20–20' Directive), which prescribes a 20 per cent reduction in primary energy consumption by the year 2020, has been dismissed by mainstream Russian experts as wishful bureaucratic thinking. In September 2010, Putin confidently predicted: 'According to most forecasts, natural gas consumption in Europe will increase by 200 billion cubic meters in the next 10 years' (Putin, 2010a). The EU guideline (set by the same Directive) of increasing the share of renewable energy sources to 20 per cent of total energy mix is seen in Moscow as a deliberate violation of market competition by granting fat subsidies to wind farms and solar panels, and has been branded as 'carbon protectionism' (Medvedev, 2010).

The EU Directive, approved in pre-crisis 2007, may indeed be off-target, because the prescribed investments in energy efficiency could prove unaffordable in today's era of severely cut state budgets (Simonov, 2011). For that matter, the International Energy Agency (IEA) has predicted that European gas consumption will increase from 508 bcm in 2009 to 593 bcm by 2020 – and that under the 'New Policies Scenario' (IEA, 2011: 159). That does not mean, however, that Russia is set to march in through the opening between the fast-falling production of and slow-rising demand for natural gas in Europe (providing the decade will not be lost to stagnation). The public obsession with 'carbon footprints' may have eased after the fiasco of the UN Climate Summit in Copenhagen in December 2009, but the reputation of Russian gas remains badly damaged by the two high-resonance interruptions of deliveries in 2006 and 2009 (Stegen, 2011). Even for the Germans, the special privileges do not quite compensate for the drawbacks of having to deal with Gazprom, which the US ambassador – according to *WikiLeaks* – at one point characterized as a grossly inefficient and corrupt monopoly (Smirnov, 2011).

The European Commission is reluctant to abandon the ambitions of its politically correct energy policy and insists that the Copenhagen climate agenda was maliciously derailed by a cabal of oil producers (including Russia) and polluters (primarily China). It may take another year or so for the politicians, who have to deal with the consequences of protracted recession, to impress upon the bureaucrats that the imperative of placing the policy on a more sound economic foundation cannot be ignored, so that the aim of increasing energy efficiency can be corrected by cost-efficiency criteria. The diversification guideline thus becomes less politically-driven and more focused on promoting competition on liberalizing markets – and that will require greater opening of national markets by, for instance, constructing the long-overdue inter-connector pipelines. European energy 'champions', including E.ON, are perhaps not enthusiastic supporters of liberalization, but they have stopped trying to fight this trend and are now working on mechanisms to use for exploiting competitive and even saturated markets to their own interest (Romanova, 2010). Moscow had generally ignored the creeping tightening of EU regulations, so it was caught unprepared by the Commission's major move against Gazprom's complex system of co-owned assets and murky practices in Europe in autumn 2012.

Gazprom had been aware of the 'hostile' deliberations in the Commission and had sought to neutralize those by striking several low-cost compromise deals with select customers on selling portions of the contracted volumes on gas spot-market prices. However, these concessions proved insufficient to prevent a large-scale probe into Gazprom's activities, focused on its preferred model of long-term contracts on fixed volumes (which cannot be re-exported) and aimed at eliminating the old pricing formula based on oil prices (see 'Burst valves', 2012). Putin accused the EU leadership of politicizing the well-established gas business and ordered Gazprom not to cooperate with the investigation, but the Commission resolutely insisted on opening all confidential bilateral contracts (Mordyushenko and Dzhodzhua, 2012). This clash, in which Gazprom stands slim chances of winning, follows the protracted lamentations in Moscow about the prescription for separating the production, transportation and supply parts of the gas business, as prescribed by the 'Third Package' (Astakhova, 2011; Melnikova and Gavrilova, 2011). Visiting Germany in November 2010, Putin lashed out against the EU legislation: 'I think mechanical application of the provisions of the Third Energy Package puts a serious question mark over the plans to modernize energy systems and develop infrastructure' (Putin, 2010b; see also Kolesnikov, 2010).

Putin's anger has been targeted mainly at the remaining uncertainties around the Nord Stream pipeline project, which after a great many acrimonious debates moved into the phase of physical construction in spring 2010, and was officially opened on 18 November 2011 (Whist, 2008; Godzimirski, 2011). The rationale for this project was constructed a decade ago; today, the fundamentals appear far from solid, as the gas for the pipeline is not yet available in necessary quantities (due primarily to the collapse of the Shtokman project); moreover, demand in the newly-saturated North European market is uncertain. There are some profits to be harvested from the construction of the supporting infrastructure in Russia, but the main purpose of this financially nonsensical pipeline appears to be to anchor Germany firmly to Russian gas platforms (Korchemkin, 2010). Both E.ON and BASF (as well as French GDF Suez and the Dutch Gasunie) have minority stakes in Nord Stream AG (Gazprom has 51 per cent) and are working on the assumption that the elimination of transit problems will make it safe to increase their imports of Russian gas, providing it is reasonably priced (Grib and Egikyan, 2010).

The undeniable truth (albeit still denied by Moscow) is that the only way to fill this pipeline with gas is by diverting part of the flow that currently goes through Belarus. This generally fits with the Russian concept of reducing transit dependency by diversifying export channels, but will inevitably increase the transport costs, because the Yamal–Europe pipeline (capacity: 33 bcm per year) offers a more economical route. The shift to the undersea route is certain to affect relations between Russia and Belarus, which are now regulated by the painstakingly negotiated deal on the common economic space but have a long record of disputes and breakdowns (Gavshina and Mazneva, 2011). Another party affected by the construction of the Nord Stream pipeline is Poland, which was pressured by Moscow to secure for Gazprom the ownership rights over the 'strategic' pipeline in exchange for a new deal on higher volumes and lower prices on exported gas (Dempsey, 2010; Roth, 2011). Energy intrigues involve also the Baltic trio (particularly the Mazeikiu oil refinery in Lithuania), but their scope has been much diminished by earlier Russian efforts at 'diversification' focused on constructing the Baltic Pipeline System (BPS), which effectively stopped export of Russian oil through Ventspils and other Baltic terminals (Sytin, 2010; Orlov and Shirokova, 2009). Nevertheless, Lithuania is taking Gazprom to the Stockholm Arbitrage, demanding 1.5 billion euro in compensation for overpriced gas (Kulikov, 2012b).

Overall, the opening of the Nord Stream pipeline (the second trunk was opened in October 2012, raising capacity to 55 bcm) will hardly help Russia in strengthening its position in the North European gas market, and is likely to pay far fewer geopolitical dividends than expected by Moscow (Shokhina, 2012). The key factor in reshaping this market remains Germany's interest in, and attitude towards, Russian gas – which could fluctuate, depending upon Moscow's behaviour in other energy markets. Nevertheless, the German decision to halt the production of nuclear energy may well have some impact on Russia's position on the German gas market (Fischer and Geden, 2011; Kosobokova and Geltishchev, 2011).

Cutting Turkey in, Ukraine out, and the Balkans through

Perhaps most precarious is the energy security situation in the southeastern quadrant of the European market. It is here that Russia has put forward the most ambitious plans and is facing difficult choices and challenges. It is also here that the consequences of the Russian–Ukrainian 'gas war' of January 2009 were the most severe. Long before the current troubles, Gazprom had aimed at expanding in this market – and it is still working on the assumption that the crisis will soon be over and there will be no need to revise its assessments of expanding demand for gas and electricity in the emerging European economies like Bulgaria or Serbia.

Here we might recall the first inflated assessment of demand that was made in the late 1990s regarding the Turkish market: this led to the costly construction of the Blue Stream undersea gas pipeline (capacity 16 bcm). For years, this pipeline remained under-utilized, and Gazprom was forced to bargain the prices down even if Moscow and Ankara were eager to portray it as a breakthrough in cooperation (Barysch, 2007). Another faulty assessment dismissed the Baku–Tbilisi–Ceyhan (BTC) pipeline as a political project with feeble economic foundations, but the consortium of Western 'majors' led by the BP took a calculated risk – and the climb in oil prices secured a healthy return on their investment (Starr and Cornell, 2005). Yet another serious miscalculation, noted above, was the decision to purchase all available gas in Central Asia at 'European' prices: this proved so costly for Gazprom that Turkmenistan was told bluntly by Moscow to take its gas elsewhere – just not to Europe.

This track record of blunders tells us not only about the quality of strategic planning in Russian energy policy but also about the consistent pattern of disregarding common economic sense and putting political expediency first. The Russian–Ukrainian 'gas war' of January 2009 constitutes the most solid evidence of this pattern. Although its impact has been countered by the economic and political tribulations of the ongoing economic crisis, the fundamental insecurity of energy supply in the complex EU–Russia–Ukraine–Turkey interface is unlikely to disappear in the near future (Baev, 2010a; Tekin and Williams, 2009). The central underlying problem is the clash of interests between Russia and Ukraine: Russia is determined to reduce gas transit by at least one half, whereas Ukraine aims at maximizing the benefits from transit.

Energy relations between these two states were complicated throughout the 1990s, but Gazprom and Naftogaz managed to ease the tensions stemming from Ukraine's need to import increasing volumes of gas at the lowest possible prices by a series of non-transparent and progressively more corrupt deals (Stern, 2005: 86–98). It was Ukraine's 'Orange Revolution' in late 2004 that escalated those tensions to the level of state conflict, when Putin made Gazprom instrumental for his strategic aim of turning the tide of 'colour revolutions' and ensuring the collapse of the 'Orange' coalition in Ukraine by pushing the country to the brink of bankruptcy (Panyushkin and Zygar, 2008). The development of the Russian–Ukrainian gas crisis in the second half of that decade has since been examined as regards every possible economic element and political prejudice; however, what is relevant here is not Putin's tactical game of punishing Ukraine with price increases but his strategic plan for diversifying export channels away from this 'brotherly neighbour'.

The main element in this plan is the South Stream gas pipeline across the Black Sea, designed as a joint project with the Italian ENI (and possibly French EDF as a minor partner) with a capacity of 30 bcm – which Putin in mid-2009 doubled with a stroke of the pen to 63 bcm (Zygar, 2009). The decisive defeat of the 'Orange' forces in Ukraine in the January 2010 elections and the swift consolidation of political control by President Viktor Yanukovich should have weakened the political imperative for Russian diversification and created a breathing space for re-examining the problematic economic rationale for the South Stream project. No such space has come to be. Yanukovich did achieve rapid if rather dubious progress in negotiating an agreement on reducing prices on imported gas in exchange for extending Russia's lease on the Sevastopol naval base, but he has so far failed in securing a commitment from Gazprom to transport the bulk of its European export through the shortest Ukrainian route.

Yanukovich's proposition, still on the table in spring 2013 boils down to forming a trilateral consortium consisting of Ukraine, Russia and the EU, to manage and modernize the gas infrastructure in Ukraine. The investments required would be at least five times lower than those entailed in constructing the South Stream; moreover, there would be the further benefit of utilizing the vast gas storage facilities in Ukraine (Ivzhenko, 2012). Gazprom has evinced scant interest in this unquestionably cost-efficient proposition, pointing out that Ukraine has remained adamant about retaining full ownership of its pipelines, and Putin prefers to hedge against the transit risk, fearing that the proposed consortium could become paralysed if in the next elections Yanukovich were to be defeated by a pro-Western politician. As for the EU, it has remained reluctant to do business with the 'post-Orange' Ukraine, and up until 2012 stuck to its own plan for diversification centred on the Nabucco pipeline project (Baev and Øverland, 2010; Fernandez, 2011). The irrefutable weakness in the EU design for the Southern 'corridor' is shortage of supply: Turkmenistan is hard to reach, Iran remains off-limits, and production in Azerbaijan (including the maturing Shah Deniz II project) is only just enough to satisfy its own, Georgian and Turkish demands, leaving only some 10 bcm to be delivered to Italy via the Trans-Adriatic Pipeline (TAP), which Statoil, EGL and E.ON Ruhrgas are set to construct (Hoedt, 2011). To all intents and purposes, a series of decisions on postponement and capacity-curtailing signified a collapse of the Nabucco project during 2012 (Riley, 2012; Simonov, 2012).

South Stream does not need any new volumes of gas (although Russia has persisted in its hard-to-refuse offers to Azerbaijan). However, it suffers from serious cost problems (Gazprom's latest estimate of 15.5 billion euro and even the unofficial target figures of 19–24 billion euro appear modest), as well as an organizational conundrum (Grib, 2011). In legal and operational terms, South Stream is not one but half a dozen successive pipelines, starting with the undersea section (a joint venture with Italian ENI), continuing with the Bulgarian, Serbian, Hungarian, Slovenian and Greek pipelines, each with different partners and specific transit agreements. Negotiations on these deals, which as yet are far from watertight, have required deep personal involvement on the part of Vladimir Putin; moreover, the highly probable solvency problems in these states mean that Russia may find itself facing a great many transit complications.

The necessary condition for advancing this project was Turkey's consent for laying the extra-long pipeline across its exclusive economic zone in the Black Sea. Ankara was firmly set on extracting maximum benefits here. Pursuing the idea of 'gas hub', which finds few supporters in the EU, Turkey has insisted on doubling the capacity of Blue Stream, thus leaving Ukraine with even thinner transit. Italian ENI has entertained ideas of expanding this pipeline even further and abandoning both South Stream and Nabucco – which would be a plan too far for both Moscow and Brussels (Grib and Gabuev, 2010).

Ankara's requests have not been limited to gas, but have included an oil pipeline from Samsun to Ceyhan, which would minimize tanker traffic through the Bosporus but also 'kill' the alternative bypass pipeline Burgas–Alexandroupolis. Moscow had invested considerable effort in striking the deal on the latter pipeline, but the project was abandoned when in December 2011 Bulgaria announced its withdrawal (RBK 2011).

The main priority for Russia now is the privately owned Caspian Pipeline Consortium (CPC), which agreed on doubling the capacity of the oil pipeline Tengiz–Novorossiisk (currently transporting 34 million tons of oil a year) (Ponomarev, 2010b). This gentle control over oil exports from Kazakhstan would not only provide Russia with a useful asset in its energy bargaining with Turkey, but would also implicitly reduce the geopolitical profile of the BTC pipeline.

The net results of Russia's hard efforts at opening new oil and gas export channels to Europe in the Black Sea area may not yet have been achieved, but seem set to entail a drastic curtailing of Ukraine's role and a significant boost to Turkey's role in providing transit. In fact, in addition to making scant economic sense, this deviation from the most cost-efficient route does not necessarily jibe with Russia's political interests in building closer ties with its most important neighbour.

Developments in 2011–2012 – with the trial of Yulia Timoshenko, where Ukraine's former prime minister was sentenced to seven years in prison for her role in signing the 2009 gas agreement with Russia, and the continued tensions in Russian–Ukrainian energy relations, with Russia rejecting any idea of abandoning construction of South Stream in order to accommodate Ukrainian interests – show how great a conundrum the Ukrainian–Russian energy relationship is. High-level Russian officials have mentioned the draft plan for constructing an LNG plant near Novorossiisk and have suggested adding another trunk to the Nord Stream pipeline (Dohmen and Jung, 2012), signalling their doubts about the South Stream, but these alternatives pursue the same dual goal – to expand exports to Europe and to reduce transit dependency on Ukraine.

Keeping China Friendly, the USA Attentive, and Central Asia Stable

Russia has every geopolitical reason to be worried about the fate of its Far East, with its dwindling population and chronically depressed economy, oriented primarily towards the booming markets in Asia–Pacific and essentially disconnected from the European part of the country. These reasons are not spelled out in the National Security Strategy or the Military Doctrine, but they are readily discernible behind the special attention that the Russian leadership has been paying to its Eastern periphery, which is visited more often than any other region, except perhaps St Petersburg. The central driver of these concerns is China and its astounding growth, which was not interrupted by the global economic crisis. This 'emerging economy' has graduated after the first decade of the new century to become a truly great power easily eclipsing Russia (Lukyanov, 2012).

Russia's answer, perhaps the only possible one, to the challenge of China's rise is to engage it in a strategic partnership, with energy export as a key element (Lo, 2012). On numerous occasions during his first presidency Putin made inflated promises, but those were remarkably slow in materializing, so Medvedev found himself in the rather awkward position of reiterating dubious pledges; and Putin must up his estimates, as at the APEC summit in Vladivostok (Likhacheva, Savelieva and Makarov, 2010; Tavrovsky, 2012). The main reason for Russia's inability to deliver on the many memoranda of understanding has been the slow development of the oil and gas fields in Eastern Siberia, where massive investments are required, while Moscow remains adamant on not allowing the CNPC or other Chinese companies to invest in upstream joint projects. If Beijing is irritated by these procrastinations, it has been politically savvy enough not to express its disappointment – but it is obviously not counting on any big increase in the incoming energy flows from the North in the near future.

Sustaining the efforts at building new energy provinces in Eastern Siberia, Moscow has been constantly agonizing over how much oil and gas export could be prudently and profitably diverted from the traditional European markets towards insatiable China (Baev, 2008; Zhiznin, 2011). Putin's current choice is to expand the capacity for shipping oil from Western Siberia to the East, while putting on hold any projects related to natural gas. This choice can be seen in the case of the East Siberia–Pacific Ocean (ESPO or VSTO) pipeline, the first trunk of which was opened in 2010. Some 15 million tons of oil were pumped through in 2011 which was about 5 per cent of China's total import, and no significant increase was registered in 2012 (Inozemtsev, 2012). The huge costs of construction are paid with the USD 10 billion loan granted to Transneft by the China Development Bank, which essentially means pre-payment for the oil that will be delivered in the next few years (Lenta.ru, 2011). One aspect of this project that was first exposed in the Russian blogosphere by the anti-corruption crusader Aleksei Navalny, and later elaborated in the mainstream media, is the multi-million overspending and falsification of expenditures. Here the current management of Transneft blames the team of its former CEO Semyon Vainshtok, who was allowed to leave for abroad on a 'golden parachute' (Gavshina and Dmitrienko, 2010).

What is politically significant about the VSTO project is that Putin has insisted on accelerated construction of the second trunk connecting the first trunk with the oil terminal in Kozmino, regardless of cost. That would make it possible to increase oil exports to Japan, Korea and other Asia–Pacific destinations, thereby avoiding an exclusive orientation to the Chinese market, inexhaustible as it might seem. This expensive diversification effort has clearly been driven by the desire to escape from the trap of dependency upon China, a country which has grown far too strong to be an equal partner, despite the reassurances given as to its friendly intentions. Delays in opening a channel for gas trade have been caused by similar concerns. Although Moscow does not see an incentive to accept a compromise on prices – assuming that in the mid-term China would probably have to undertake a drastic shift in its pattern of energy consumption from coal to gas for environmental reasons, which would put heavy upward pressure on global gas prices (Pravosudov, 2010) – it monitors with increasing concern Chinese investments in 'unconventional' gas technologies and must take into account the risk of another 'shale gas revolution' in the making (Skorlygina, 2012).

In the meantime, Moscow has remained neutrally positive towards China's penetration into the Caspian energy province, expecting the newly-opened pipeline from Turkmenistan to render irrelevant the EU-promoted plans for a Trans-Caspian pipeline (Aliev, 2011). An equally important consideration involves China's rising stakes in preserving stability in Central Asia, particularly after the violent crisis in Kyrgyzstan in spring 2010, which revealed the limits of Russia's power-projection capabilities. Moscow has since the turn of the millennium viewed this region as an object of US geopolitical aspirations, and has sought to check such encroachments, but now the gravest risks are seen in the sudden failure of governance in fragile states like Tajikistan, Turkmenistan and Uzbekistan. Partnership with China, as institutionalized in the Shanghai Cooperation Organization (SCO), is perceived as the best available way to neutralize this risk, even if it involves the possibility of eroding Russia's own influence, for instance in the energy sphere (Blank, 2010). Thus, at the SCO summit in Beijing in June 2012, China and Turkmenistan agreed to raise the target figure for volumes to be exported by the end of the decade to 65 bcm, while a plan for a new pipeline crossing northern Afghanistan was also discussed (Socor, 2012). There is also the growing recognition that the USA has in fact been playing a stabilizing role in the region, with its interests not energy-focused but centred on securing transit to Afghanistan. Moscow further hopes that its strengthened energy profile in Asia–Pacific – not limited to the humble role of an 'appendage' of China – could constitute a major asset in relations with Washington, since the northern part of this vast region is becoming a main focus of US foreign and security policy.

The sum of Russian efforts in expanding the flows of oil and gas eastwardly is far less impressive than what the rhetoric of diversification seeks to talk into existence. The production bases in Eastern Siberia and the Far East are underdeveloped, the redirection of oil export to China is not cost-efficient, and Russia's gas policy remains essentially Euro-centric. The only way to alter this pattern is by organizing joint ventures with Western (or for that matter Eastern) energy 'majors' for developing large-scale projects, like the Kovykta gasfield. Although Moscow seemed reluctant in 2010 to relax its counterproductive state control, in 2011–2012 several joint projects were launched involving Russian (primarily Rosneft) as well as Western companies. This may signal the opening of a new chapter in foreign companies engagement in Russia (Overland et al., 2012; Kogtev and Lukina, 2012).

Conclusions

Logically, the pronounced expansion of state ownership of, and control over, Russia's oil and gas industries should have led to improved capacities for strategic planning in the energy sector, where a key guideline is diversification of export. In reality, however, the various diversification projects have advanced in uncoordinated surges driven by political expediency, poor personal judgement and corrupt incentives. This chaotic planning has been further muddled by the impacts of the domestic recession and shifts in global energy markets that are more often denied rather than properly evaluated. The new speculative bubble in the oil market is seen as a restoration of 'natural' growth, and the oversupply on the natural gas market is dismissed as an aberration (Khaitun, 2012).

Pipeline projects like the BPS and Nord Stream in the northwest, the CPC and Blue Stream/South Stream in the southwest, and the VSTO in the east are due to be completed or expanded by 2015. This prioritization seems set to create a significant free export capacity because the development of 'greenfields' has been lagging behind, due to sustained under-investment in the upstream. In the oil sector, this free capacity can be partly utilized for transporting new oil from Kazakhstan as the Kashagan project comes online. By contrast, in the gas sector, plans for bringing new volumes from Turkmenistan via the new Pri-Caspian pipeline have been practically abandoned, and Azerbaijan is to send only symbolic amounts of its new production northwards – so the free capacity will be used for achieving greater transit security by switching from one pipeline to another.

Besides being hugely expensive, putting pressure on transit states in order to reduce transport costs and increase export prices is not necessarily in the long-term political interest of Russia. Oil and gas quarrels have bedevilled the ambitious plan for building a Union state with Belarus, but it is the pattern of relations with Ukraine that is likely to suffer the most serious damage. Although the Yanukovich government cannot be described as directly pro-Russian, it has certainly been placing major emphasis on improving relations with Russia, and on optimizing gas relations in particular. Yanukovich's big plan is to engage both Russia and the EU in modernizing the gas infrastructure of Ukraine, with the goal of raising the transit charges and reducing the import prices. Moscow's stubborn promotion of the South Stream project undermines this plan to a far greater degree than the EU support for the Nabucco project (or what is left in the Southern corridor). This diversification may result not only in massive budget deficits but also a self-perpetuating economic recession in Ukraine – and Moscow cannot possibly wish to see the failure of this paramount neighbour state with, at least for the time being, a rather Russia-friendly government.

As it is argued in many contributions to this volume, the aims of Russia's energy policy (to the degree they can be discerned) correspond poorly with the urgent need to modernize the country: indeed, implementation of these aims often amounts to counter-modernization. Russia's chief aim in diversifying energy exports (oil in particular) towards the Asia–Pacific is to build a solid foundation for its strategic partnership with China, but since the development of reserves in East Siberia is proceeding slowly and remains closed for international involvement, Beijing is still doubtful about Moscow's real intentions. The main Russian goal for diversifying the channels of export, first of all of natural gas, in a westerly direction is to capture a larger share of the European market, but the political agenda behind these costly efforts has made the EU reluctant to accept greater dependency on Gazprom supply, and has derailed the commitment to partnership in modernization. Russia may soon have greater flexibility in manoeuvring between different markets and transit routes, but that will hardly boost its own eroding energy security – and it is likely to come at an almost prohibitive price.

References

Aleksashenko, S. 2012. Russia's economic agenda to 2020. *International Affairs*, 88(1), 31–48.

Aliev, A. 2009. Zheltyi potok [The yellow stream]. *Expert*, 15 December.

Aliev, A. 2011. Soderzhanie gaza [Gas content]. *Expert*, 14 January.

Astakhova, A. 2011. 'Gazprom' vylezaet iz okopa. Monopolistu pridetsya smiritsya s novymi evropeiskimi pravilami [Gazprom is coming out of the trenches. The monopolist has to cope with new European laws]. *Nezavisimaya Gazeta NG-Energia*, 8 November.

Baev, P. K. 2008. Asia-Pacific and LNG: the lure of new markets, in K. Barysch (ed.), *Pipeline, Politics and Power: The Future of EU–Russia Energy Relations*. London: Centre for European Reform (CER), 83–92.

Baev, P. K. 2010a. Energy intrigues on the EU's southern flank: applying game theory. *Problems of Post-Communism*, 57(3), 11–22.

Baev, P. K. 2010b. Russia abandons the 'energy super-power' idea but lacks energy for modernization. *Strategic Analysis*, 34(6), 885–96.

Baev, P. K. 2010c. Virtual geopolitics in Central Asia: U.S.–Russian cooperation vs. conflict of interests. *Security Index*, 16(1), 29–36.

Baev, P. K. and I. Øverland. 2010. The South Stream versus Nabucco pipeline race. *International Affairs*, 86(5), 1075–90.

Barysch, K. 2007. *Turkey's role in European energy security* [Online: Centre for European Reform]. Available at http://www.setav.org/ups/dosya/24478.pdf [accessed 15 November 2012].

Blank, S. 2010. A Sino-Uzbek axis in Central Asia? [Online: *CACI Analyst*, 1.] Available at http://www.cacianalyst.org/?q=node/5395 [accessed 15 November 2012].

Bradshaw, M. 2008. The Sakhalin saga. *Soundings*, 40 (Winter), 56–68.

Burst valves 2012. *The Economist*, 15 September.

Dempsey, J. 2010. Europe seeks to block Polish gas contract. *New York Times*, 10 October.

Dohmen, F. and A. Jung. 2012. Gazprom hopes to build second Baltic Sea pipeline. *Spiegel International*, 18 May.

Fernandez, R. 2011. Nabucco and the Russian gas strategy vis-a-vis Europe. *Post-Communist Economies*, 23(1), 69–85.

Fischer, S. and O. Geden. 2011. Europeanising the German Energy Transition, *SWP Comments* 33, November. Berlin: SWP.

Gavshina, O. and D. Dmitrienko. 2010. Navalny trebuet privlech eks-menedzherov Transnefti k ugolovnoi otvetstvennosti [Navalny calls for criminal investigation of the former Transneft management]. *Vedomosti*, 16 November.

Gavshina, O. and E. Mazneva. 2011. Neft zapakhla gazom [Oil smells like gas]. *Vedomosti*, 12 January.

Godzimirski, J. M. 2011. Nord Stream: globalization in the pipeline, in E. W. Rowe and J. Wilhelmsen (eds), *Russia's Encounter with Globalization*. London: Palgrave Macmillan, 159–84.

Grib, N. 2010. Yuzhnyi potok priravnyali k Nabukko [South Stream is made equal to Nabucco]. *Kommersant*, 1 December.

Grib, N. and A. Gabuev. 2010. Rossiya blokiruet nedruzhestvennoe sliyanie [Russia blocks an unfriendly merger]. *Kommersant*, 16 March.

Grib, N., S. Egikyan and E. Kiseleva. 2010. Nemtsy sdayut Gazprom [Germans are dumping Gazprom]. *Kommersant*, 24 November.

Grib, N. and S. Egikyan. 2010. Nord Stream napolnilsya do predela [Nord Stream is full to the limit]. *Kommersant*, 11 August.

Guriev, S. and A. Tsyvinski. 2010. Challenges facing the Russian economy after the crisis, in A. Åslund, S. Guriev and A. Kuchins (eds), *Russia after the Global Economic Crisis*. Washington, DC: Peterson Institute for International Economics, 9–39.

Hoedt, R. ten. 2011. To us it is just a money game. *European Energy Review* [Online: European Energy Review] Available at: http://www.europeanenergyreview.eu/index.php?id=2638 [accessed 15 November 2012].

IEA 2009. *World Energy Outlook 2009*. Paris: International Energy Agency.

IEA 2010. *World Energy Outlook 2010*. Paris: International Energy Agency.

IEA 2011. *World Energy Outlook 2010*. Paris: International Energy Agency.

Inozemtsev, V. 2012. Na obochine velikogo okeana [On the verge of a great ocean]. *Ogonyok*, 3 September.

Ivzhenko, T. 2012. Gazovyi krizis vne grafika [Out of schedule gas crisis]. *Nezavisimaya gazeta*, 18 September.

Khaitun, A. 2012. Gazprom na glinyanykh nogakh [Gazprom on feet of clay]. *Nezavisimaya gazeta, NG-Energiya*, 9 October.

Khodov, I. 2010. Mirovoy krizis i rossiyskiy gaz. Eksport rossiyskogo gaza w Germaniyu, dolgosrochnyye faktory padeniya sprosa[World crisis and Russian gas. Export of Russian gas to Germany, long-term factors in the decline of demand]. *Svobodnaya Mysl* (12), 91–106.

Kogtev, Y. and V. Lukina. 2012. Partnersky goskapitalizm [Partner state-capitalism]. *Neft i Gaz* in *Kommersant*, 30 August.

Kolesnikov, A. 2010. Vladimir Putin provel razboi poletov [Vladimir Putin on Russian business in Europe]. *Kommersant*, 27 November.

Korchemkin, M. 2010. Gazprom multiplies the costs of pipelines construction. *Forbes.ru*, 25 November.

Kosobokova, T. and P. Geltishchev. 2011. Dmitriy Medvedev obeshchayet nemtsam rossiyskuyu alternativu AES [Dmitriy Medvedev promises Russian alternative AES to Germans]. *RBC Daily*, 19 July, available at: http://www.rbcdaily.ru/2011/07/19/focus/562949980670174 [accessed 15 November 2012].

Krysek, T. F. 2007. Agreements from another era: production sharing agreements in Putin's Russia. Oxford: Oxford Institute for Energy Studies.

Kulikov, S. 2012a. Evropa davit na Gazprom [Europe puts pressure on Gazprom]. *Nezavisimaya gazeta*, 7 September.

Kulikov, S. 2012b. Litva tyanet Gazprom v sud [Lithuania pulls Gazprom to the court]. *Nezavisimaya gazeta*, 4 October.

Kupchinsky, R. 2009. *Russian LNG – the future geopolitical battleground.* Washington, DC: Jamestown Foundation.

Lenta.ru 2011. *Transneft has spent Chinese loan on the VSTO.* [Online: Lenta.ru]. Available at: http://lenta.ru/news/2011/01/14/credit/ [accessed 15 November 2012].

Likhacheva, A., A. Savelieva and I. Makarov. 2010. Daily bread and water. *Russia in Global Affairs*, July/September (http://eng.globalaffairs.ru/number/Daily-Bread-and-Water-15003).

Lo, B. 2012. A Partnership of Convenience. *International Herald Tribune*, 7 June.

Lukyanov, F. 2012. Na gegemona ne tyanet [Hegemon does not appeal] [online: Gazeta.ru]. Available at: http://www.gazeta.ru/column/lukyanov/4616101.shtml [accessed 7 July 2012].

Makarkin, A. 2011. The Russian social contract and regime legitimacy. *International Affairs*, 87(6), 1459–75.

Medvedev, D. 2010. *Remarks at the meeting of the Russian Security Council on 17 March 2010.* Available at: http://eng.kremlin.ru/speeches/2010/03/17/1931_type82913_224806.shtml [accessed 15 June 2012].

Melnikova, S. and Gavrilova, Y. 2011. Prinyat nelzya otkazatsia [Cannot deny acceptance]. *Nezavisimaya Gazeta*, 13 April.

Milov, V. 2006. Neo-Cons plans and sober reality. *Russia in Global Affairs*, October/December (http://eng.globalaffairs.ru/number/n_7340).

Mitrova, T. 2011. Future development of LNG in Russia [online: Institute of Enegy Research and Russian Academy of Science] http://www.eriras.ru/images/papers/mitroval.ppt [accessed 15 June 2012].

Mordyushenko, O. and T. Dzhodzhua 2012. EC nashel klyuch k kontraktam Gazproma [EU found keys to Gazprom contracts]. *Kommersant*, 5 October.

Naumov, I. 2011. Kovykta obrela novogo hozyaina [Kovytka has a new owner]. *Nezavisimaya gazeta*, 2 March.

O'Hara, S. and L. Hongyi 2011. China's 'dash for gas': challenges and potential impacts on global markets. *Eurasian Geography and Economics*, 52(4), 501–22.

Orlov, D. and E. Shirokova 2009. Oil transit: another Western vector. *Nezavisimaya gazeta – NG Energiya*, 2 September.

Overland, I., H. Kjaernet and A. Kendall-Taylor 2009. *Caspian Energy Politics: Azerbaijan, Kazakhstan and Turkmenistan*. Central Asian Studies. London: Routledge.

Overland, I., J. Godzimirski, P. L. Lunden and D. Fjærtoft 2013. Rosneft's offshore partnerships. The re-opening of the Russian petroleum frontier? *Polar Record*, 49(2), 140–153.

Panyushkin, V. and M. Zygar 2008. *Gazprom: Novoe Russkoe Oruzhie* [Gazprom: new Russian weapon]. Moscow: Zakharov.

Ponomarev, V. 2010a. Gotovnost no. 2 [Readiness no. 2]. *Expert*, 15 December.

Ponomarev, V. 2010b. Vremya truby [The time of the pipe]. *Expert*, 25 August.

Pravosudov, S. 2010. Obrechennost na sotrudnichestvo [Pre-determined cooperation]. *Nezavisimaya gazeta – NG Energiya*, 14 December.

Putin, V. 2006. Transcript of meeting with participants in the third meeting of the valdai discussion club [online: Kremlin Archives]. Available at: http://archive. kremlin.ru/eng/speeches/2006/09/09/1209_type82917type84779_111165. shtml [accessed 15 June 2012].

Putin, V. 2010a. Speech during Prime Minister's visit to the Nord Stream construction site. 20 September [online: Official Prime Minister page]. Available at: http://premier.gov.ru/eng/visits/ru/12265/events/12280/ [accessed 15 June 2012].

Putin, V. 2010b. Speech at the meeting with managers of major German companies 26 November [online: Official Prime Minister Page]. Available at: http://premier.gov.ru/eng/visits/world/13103/events/13118/ [accessed 15 June 2012].

Putin, V. 2012. O nashih ekonomicheskih zadachah [About our economic tasks]. *Vedomosti*, 30 January.

Rahr, A. 2007. Germany and Russia: a special relationship. *Washington Quarterly*, 30(2), 137–46.

RBK 2011. *Bolgariya prinyala resheniye otkazatsya ot proyekta 'Burgas – Aleksandrupolis'*. [Bulgaria decides to cancel the Burgas-Alexandroupolis project] Available at: http://www.rbc.ru/rbcfreenews/20111207152002.shtml [accessed 15 June 2012].

Riley, A. 2012. There is life for the southern corridor after Nabucco. *European Energy Review*, 12 March.

Romanova, T. 2010. Energy security without panic: Russia–EU energy dialogue moving back to economy. *Russia in Global Affairs* [online] 9(2). Available at: http://eng.globalaffairs.ru/number/Energy_Security_Without_Panic-14900 [accessed 15 June 2012].

Rossiyskaya Gazeta 2011. Rossiya i Kitay soglasovali formulu tseny na gaz [Russia and China agreed on a price formula for gas]. 27 September.

Roth, M. 2011. Poland as a policy entrepreneur in European external energy policy: towards greater energy solidarity vis-à-vis Russia? *Geopolitics*, 16(3), 600–625.

Russian Government 2009. *Energeticheskaya strategiya Rossii na period do 2030 goda [Energy Strategy of Russia through 2030]*. Adopted by the Decree of the Government of RF # 1715-p, 13 November 2009 ed. Moscow: Government of the Russian Federation.

Satayev, D. 2010. Brak po raschetu [Marriage by calculation]. *Nezavisimaya gazeta*, 26 October.

Shokhina, Y. 2012. Podarok ot Millera [Present from Miller]. *Expert*, 8 October.

Simonov, K. 2011. Gazovyj rynok Evropy. Krakh industrii energeticheskikh prognozov [European gas market. Crash of energy prognosis industry]. *Nezavisimaya Gazeta NG-Energia*, 13 September.

Simonov, K. 2012. Deshevy gaz i 'myshelovka' [Cheap gas and 'mousetraps']. *Expert*, 29 June.

Skorlygina, N. 2012. Kitai prevratit ugol v gaz [China will turn coal into gas]. *Kommersant*, 8 October.

Smirnov, S. 2010. Gazovyi razvod [Gas divorce]. *Expert-Kazakhstan*, 25 January.

Smirnov, S. 2011. Gazprom – korrumpirovannaya nerynochnaya rentnaya monopoliya [Gazprom – corrupt rent-seeking monopoly]. *Vedomosti*, 6 January.

Socor, V. 2012. Beijing proposes Turkmenistan-China pipeline through northern Afghanistan. *Eurasia Daily Monitor*, 19 June.

Starr, F. and S. Cornell 2005. *The Baku–Tbilisi–Ceyhan Pipeline: Oil Window to the West*. Washington, DC: Johns Hopkins University.

Stegen, K. S. 2011. Deconstructing the 'energy weapon': Russia's threat to Europe as case study. *Energy Policy*, 39, 6505–13.

Stent, A. 2007. Berlin's Russia challenge. *The National Interest*, March/April [online: National Interest] http://nationalinterest.org/article/berlins-russia-challenge-1479 [accessed 15 June 2012].

Stern, J. 1999. The origins and evolution of Gazprom's export strategy, in R. Mabro and I. Wybrew-Bond (eds). *Gas to Europe: The Strategies of the Four Major Suppliers*. Oxford: Oxford University Press, 135–200.

Stern, J. 2005. *The Future of Russian Gas and Gazprom*. Oxford: Oxford University Press.

Stern, J. 2010. European gas: competitiveness, security, carbon reduction. Presentation at the Statoil autumn conference. Available at: http://www.statoil.com/en/NewsAndMedia/Calendar/Downloads/Presentationper cent20byJonathan per cent20Stern.pdf [accessed 15 November 2012].

Sytin, A. 2010. Vokrug tranzita. Strany Baltii v borbe za 'energeticheskyyu nezavisimost' [Around transit. Baltic states in a quest of energy independence]. *Svobodnaya Mysl*, (3), 31–44.

Tavrovsky, Y. 2012. Razvorot na Vostok [Turn to the east]. *Nezavisimaya gazeta*, 6 June.

Tekin, A. and P. Williams 2009. EU–Russian relations and Turkey's role as an energy corridor. *Europe-Asia Studies*, 61(2), 337–56.

Tsyvinski, A. and S. Guriev 2010. That 70's show in Russia, available at: http://www.project-syndicate.org/commentary/tsyvinski4/English [accessed 15 June 2012].

Whist, B. S. 2008. Nord stream: not just a pipeline [online: FNI Report]. Available at: http://www.fni.no/doc&pdf/FNI-R1508.pdf [accessed 15 June 2012].

Wong, E. 2011. China quietly extends footprints into central Asia, *New York Times*, 2 January.

Zhiznin, S. 2011. Vostochnoe napravlenie. Perspektivy energeticheskogo sotrudnichestva Rossii i Kitaya [Eastwards. Prospects of energy cooperation between Russian and China]. *Nezavisimaya Gazeta NG-Energia*, 13 April.

Zygar, M. 2009. Voina potokov [The war of streams]. *Kommersant-Vlast*, 18 May.

Chapter 7

The Impact of Domestic Gas Price Reform on Russian Gas Exports

Lars Petter Lunden and Daniel Fjærtoft

Introduction

Domestic gas price reform is regarded as necessary in order to secure additional gas volumes to Europe – and Europe is, as shown in the chapters by Mitrova, Busygina and Filippov, and in Baev, the most important market for Russian gas. Gas prices in Russia are regulated at low levels, which leads to overconsumption and, indirectly, lack of necessary investment in new production capacity. Combined with the on-going decline in core West Siberian production assets, fears have arisen that Russia may have to sacrifice its exports to Europe in favour of domestic security of supply, defined by the national political leadership as the major task of the country's gas companies.

It has been argued that higher domestic prices would curb demand, through substitution effects such as fuel switching and energy efficiency, as well as income effects (lower consumption due to reduced real income). Reducing the share of domestic consumption in total production would free up volumes for export and boost revenues. Higher prices would also provide incentives for new field developments, compensating for production decline in existing fields, or even resulting in increased production.

However, the linkage between higher domestic prices and increased gas exports remains far from proven. Russia's gas export is influenced by many factors – including supply and demand in foreign markets, conditions in the regulated domestic market, and the interconnection between foreign and domestic markets. By examining the prerequisites for reduced demand, new production and Gazprom's response to higher domestic prices, this chapter investigates whether increased domestic gas prices will lead to increased Russian export. Our analysis indicates that unless domestic price hikes can be accompanied by reforms in other areas, the additional volumes of gas available for export will be limited. Price reform might also have a side-effect: it could offer further incentives for Russia to use gas exports as a tool of foreign policy.

This chapter analyses potential developments in the Russian gas sector using a simple micro-economic framework. By investigating the incentives of Gazprom and other players in the Russian gas market, conclusions can be drawn without having to take into consideration the political and personal motivations that interfere with business economics in Russia to a greater degree than in the West. Coupling this chapter with the more holistic approaches followed in other chapters should provide a deeper understanding of the dynamics of the Russian gas sector.

The study is structured around four questions, where the first three are directly related to the economics of the Russian gas sector while the last takes into account the assumed symbiosis between Gazprom and the foreign policy objectives of the Kremlin.

1. Will gas price reform reduce domestic demand, turning consumption savings into a source of export supply?
2. Will the reform have an impact on investments in gas field developments, resulting in higher gas production?
3. What incentives does the gas price reform in its current form give for Gazprom to re-allocate gas supplies?
4. Will the gas price reform provide new incentives for using gas as a tool of foreign policy?

Providing adequate answers to these questions is a challenging task indeed. The literature is rife with studies of various aspects of gas price reform and gas exports to Europe. However, our aim here is not to launch a new, in-depth investigation into each of these questions, but to identify and bring together the most important variables that affect Russian gas export capacities and decisions. By providing an overview of the dynamics of gas export, we can examine the impact of the gas price reform on Russian gas exports.

Russian Gas Price Reform

Domestic gas prices in Russia have been low compared to those elsewhere in Europe. Indeed, during Soviet times, gas was considered a public good. As can be seen in Figure 7.1, from 2003 to 2008 the gap between European and Russian gas prices increased steadily. The artificially low gas price has had several negative consequences for Russia: distorted fuel switching, a lack of investments in modern power generation facilities and other energy-efficiency measures, lost revenues to Gazprom and the tax authorities, and a substantial lack of investments in new fields and pipelines.

By 2006, the gap between domestic and foreign prices had passed 250 USD per thousand cubic meters (tcm). From 2005 to 2006, domestic consumption grew by about 6 per cent,[1] leading to heated discussions in Russia and abroad about Russia's demand outlook and future ability to meet export commitments to Europe without firm domestic action on increasing gas prices (Sagen and Tsygankova, 2008; Ahrend and Tompson, 2004). Finally, the authorities launched a price reform aimed at more effective use of gas in Russia.

---

1   Econ Pöyry Gas Database.

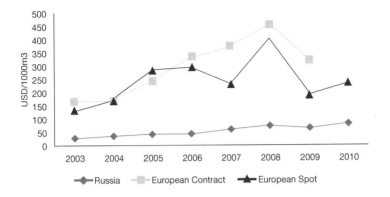

Figure 7.1   Development of European vs. Russian gas prices[2]

The reform was devised so that prices for industrial consumers would reach netback[3] parity by 2011 (Mitrova, 2009: 36). Residential consumers were exempted from the reform for the time being. This plan was approved by the government in 2007 (Decree No. 333). Gas prices were to be regulated by the Federal Tariff Service (FTS) using a formula based on the European gas price minus the cost of transport to Europe. The decree prescribed annual percentage price hikes of up to 2011. To prepare the market for the new system, 'would-be' prices based on the formula were posted on the FTS website. Since 2006, implementation of the price reform has been revised repeatedly (Kutschera and Overland, 2009: 8–9). The very high gas prices in 2008 and the financial crisis the following year had a strong impact. For example, the projected increase that was made in 2007 of a 30 per cent increase in 2010 was in 2009 revised down to 15 per cent (Hunter, 2010: 9). And in May 2008, the timing of netback parity was postponed to 2014–15, to avoid 'raising prices by 100 percent' (Fjærtoft, 2010: 8).

As of 2013, it was not clear when the gas price could reach netback parity. In fact, the prescribed interim scheme fails to provide for a seamless transition to pricing based on the netback formula. The planned netback price is linked to the average price achieved by Gazprom in Europe, which fluctuates as a result of oil-indexed contracts and spot prices. Therefore, as long as Russian reform price goals are formulated as percentage price increases in the regulated prices, it will be impossible to determine exactly when the reform will be completed. On the other hand, the Russian authorities might decide that the price formula would come into effect in say, 2014 – but that would entail loss of control over price development, as European prices are likely to keep fluctuating.

---

2   http://www.fstrf.ru/tariffs/analit_info and ICIS Heren Gas Prices.

3   In this chapter, 'netback price' concerns export prices less transport costs, taxes and import duties. Other authors sometimes define netback prices more narrowly – for example, as price less transportation costs.

Progress toward higher domestic prices has been slow compared to what would have been required to catch up to the would-be netback prices. In 2009 and 2010 the gap between foreign and domestic prices closed, but this was due mainly to contractions in European prices. At the outset of 2011 the Russian–European price gap was back where it started in 2007. Moreover, the financial crisis led to a substantial drop in European demand for Russian gas. This may have helped to assuage concerns about an immediate default on export commitments, but, more importantly, it seems to have taken some of the steam out of the reform process. The much-debated price formula has received little attention since the outcry that Russian prices might exceed average European prices due the formula's lagged link to Gazprom's predominantly oil-indexed contracts. Reform seems to have been put aside due to the effects of the financial crisis on the Russian economy, and will have to be picked up again. It remains to be seen whether an increasing price gap will fuel demands for domestic price reform. Taking into account the regime's wariness to civil discontent, as argued by Øverland and Kutschera (2009) as well as by Andresen (2008), we would venture to predict that reform progress will be slow even if it receives renewed attention. The departure of Finance Minister Kudrin in September 2011, the wave of social protests in the wake of State Duma elections in December 2011 and the relatively low support for Putin on the eve of presidential elections in March 2012 may make implementation of the reform even more daunting – and thus less probable in the current political situation.

Will Domestic Demand Reductions Increase Russian Export Supply?

Schoolbook economic theory holds that the effect of an increase in gas prices will be reduced gas demand. Everything else being equal, that should increase the amount of gas available for export to Europe or other destinations. However, this effect, as well as the degree to which it will be substantial, hinges on certain assumptions, such as sufficient price elasticity of demand. The Russian market does not determine the intersection of supply and demand. Instead, the government defines a quantity and a price, and Gazprom is obliged to provide those volumes at the prices determined by the government. If there is mutual interest, some of this production may be supplied by other producers.

The price elasticity effect is presented in Figure 7.2, with supply elasticities subdued to emphasize demand-side effects. Total supply of gas (Gazprom plus independent producers, including shares between the two) is fixed at Q_T. Supply to the domestic market is set to equal domestic demand D at prevailing prices P. Two possible demand curves are indicated: one relatively elastic, D_1; and one relatively inelastic, D_2. Exports are defined as Q_T less Q_D, domestic consumption.

Suppose current supply is defined by the intersection D_1 and P_1, giving a total allocation to the domestic market of Q_{D1}. Exports equal Q_T–Q_{D1}. A government-led price increase from P_1 to P_2 would, assuming that D_1 is true, lead to a domestic demand reduction from Q_{D1} to Q_{D2}. Exports increase up to the limit Q_{D2} – Q_{D1}. If, on the other hand, domestic demand is given by D_2, which is a more price-inelastic demand curve, the same price increase would lead to smaller demand contraction, freeing up lower volumes for export as Q_{D3} – Q_{D1} < Q_{D2} – Q_{D1}. Hence, the degree to which a given price increase may increase export supply will depend on the size of the price elasticity of domestic demand.

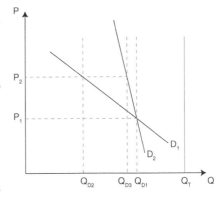

Figure 7.2   Russian demand and price dynamics

Empirical evidence on price elasticity in the Russian market is scarce. Many studies that have sought to calculate various effects of gas price increases have ended up assuming elasticity levels based on experience from other countries (see Sagen and Tsygankova, 2006; Christie, 2010). It would seem that price increases have been too few and too limited to support meaningful estimates.

It could be useful to look at the correlation between prices and domestic consumption shown in Figure 7.3. Gas prices are based on average prices received by Gazprom on the domestic market in the period 2003–2008 (Gazprom, 2010). These are plotted on the right axis in Figure 7.3. We note the positive correlation between gas prices and consumption until 2008. This observation is supported by Tatyana Mitrova (2009: 42), who finds a positive correlation between gas consumption and prices in the years between 1999 and 2006.

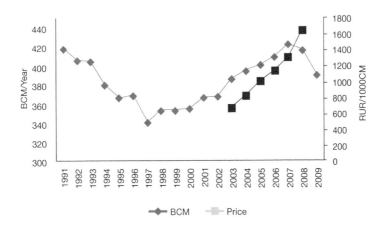

Figure 7.3   Russian gas consumption 1991–2009 (BP, 2010 and Pöyry)

GDP Matters

As yet, limited and negative price elasticity, i.e., the extent to which an increase in gas prices reduces consumption, seems to have been outweighed by a more substantial, positive income elasticity of demand – that is, the effect of increased overall income has led to an increase in gas consumption even though gas prices have increased. Figure 7.4 gives an overview of consumption patterns in Russia for the years 1995–2009. Gas consumption patterns accompanied GDP decline in the 1990s before turning to steady growth in the early 2000s, until the financial crisis took its toll also on Russian domestic consumption. Soviet-era consumption levels were re-achieved in 2007, when domestic consumption reached 422 billion cubic meters (bcm) per year. However, in 2009 domestic consumption dropped back to 390 bcm. The correlation between gas prices and GDP growth is supported by Mitrova (2009: 17).

Igor Bashmakov of the Russian Centre for Energy Efficiency also notes the importance of GDP growth on gas demand. He even argues that domestic GDP growth will be so strong that there will not be any domestic demand reduction, as prices increase according to the projected path (Bashmakov, 2005).

If the relative strength of these two effects remains constant, domestic price increases might serve only to limit demand compared to what it might have been, rather than achieving absolute reductions.

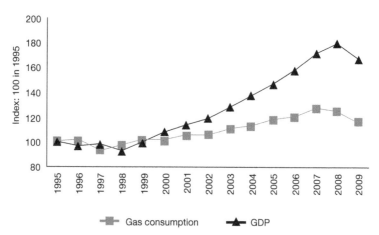

Figure 7.4   Gas consumption and GDP (2008 prices). Index: 1995=100
Source: Rosstat and Gazprom 2010.

Why is Price Elasticity Limited?

The reaction of gas demand to a price change, as for any given good, may be split into a substitution effect and an income effect. When the price rises, consumers and firms switch to alternative goods and inputs in order to minimize the cost of the desired utility or production level. However, since all gas consumption cannot be substituted, the increased spending on remaining gas usage reduces total consumption of goods as if total income had been reduced, yielding the income effect of a price increase.

Household Sector: Limited Income Effects

Low contribution of gas to the cost of the average consumer bundle may limit the income effects of price increases, because price increases will have relatively little effect on consumer budget constraints even if consumption volumes remain unchanged.

Uncertainty exists as to whether the cost of gas consumption is high compared to household income in Russia. According to the Energy Research Institute of the Russian Academy of Science, the share of total payments for electricity, heat and gas in personal incomes was 1.5 per cent in 2005, with an expected increase to 3 per cent in 2011 (Mitrova, 2009: 48). This would leave Russian consumers with one of the world's lowest ratios of gas bills to disposable income, on par with that of the USA. On the other hand, according to Kutschera and Øverland (2009), costs for water supply and sanitation become a strain on household economies if they exceed 3–5 per cent of income: that means that if the expected increase occurs, gas prices could approach levels that will unleash noticeable income effects.

A prerequisite for the income effect is that consumers can react to price signals. However, Russian consumers today often do not know how much gas they consume, and their options for adjusting consumption are limited. It is common to pay a utility fee that is independent of gas consumption; in many residential areas, the heating system is centralized for either a building or even a whole area. The only way to regulate indoor temperature is often to open the windows. Individual metering is currently being introduced, and that may make people more aware and enable them to save more energy – but progress in that field has been slow (Kutschera and Overland, 2009; Lelushkin, 2011).

Substituting Gas Consumption – Energy Efficiency and Fuel Switching

In addition to income effects, potential substitution effects play an important role in determining the price elasticity of Russian gas demand. The substitution effects are believed to be most potent, although the potential has yet to come fully into play.

Energy Efficiency – Russia's Largest Source of Supply?

In a 2008 study, the World Bank Group estimated the energy savings potential in Russia to be a whopping 45 per cent of its total primary energy consumption (Sargsyan and Gorbatenko, 2008). The volumes of gas that could be saved were equally astonishing: 240 bcm per year.

Housing and power generation stand out as the two sectors with the greatest technically and economically viable potential for savings. In sum, the technical potential of these two sectors alone amounts to approximately one sixth of Russia's total primary energy consumption. However, these are also the sectors with the lowest *financially viable* potential, which would indicate substantial benefits from increased prices.

The World Bank survey (Sargsyan and Gorbatenko, 2008) also highlights a paradox: energy-efficiency investments that would generate an attractive rate of return, even at current energy prices, are not necessarily being made. This seems to indicate the existence of administrative and informational barriers. It follows that higher prices alone will probably not unleash the full savings potential from raising natural gas prices.

Persistent delays in the gas price reform have undermined the government's credibility among investors and hence also the potential of the reform to provide suitable incentives and minimize adjustment costs. The schedule of gas price hikes has already been changed so many times that it is anyone's guess at what rate and at which level prices will end. That also means that a company considering an investment that is not profitable today but that may be profitable if netback pricing is introduced, is exposed to considerable risk regarding the timing. If the company invests too early, it will experience costly capital expenditures that will not pay off until gas prices rise. On the other hand, if it invests too late, it will have to cope with high costs due to high gas prices and high gas consumption, while simultaneously bearing large capital outlays to invest in new equipment. As yet, inertia has prevailed, and progress on energy efficiency measures has been slow.

Fuel Switching in the Wake of Higher Gas Prices

In 2008, 44 per cent of Gazprom's sales were accounted for by energy production and utility packages (Gazprom, 2010). In addition to reducing gas consumption through efficiency measures like turbine renewal and refurbishment of district heating infrastructure, providers may switch to other fuels by investing in coal-fired generation, hydro or nuclear power. Higher gas prices would increase the relative competitiveness of these fuels compared to gas, logically leading to a lower share for gas in Russia's total energy consumption.

As for coal, its ability to compete with gas should not be overestimated. In Russia, coal must be transported over substantial distances to the areas where it is to be used, which limits its competitiveness even at full netback pricing.

Other sources of energy such as nuclear, hydropower and other renewables are alternatives to gas for power and heat production. In October 2006, a USD 55 billion nuclear development programme was launched in Russia. This programme has recommended an increase in the share of nuclear energy in total production from 15.6 per cent to 18.6 per cent by 2020. Increased nuclear capacity could help reduce the contribution of gas to base-load generation and hence also consumption (Oxenstierna, 2010). Hydropower currently accounts for about 20 per cent of electricity production in Russia (RusHydro, n.d.). However, Russia has the second largest hydropower potential in the world, of which only one fifth is currently utilized. Hydropower could play an important role in replacing gas also as a swing producer in the power market. However, as with nuclear projects, hydropower projects require large investments and long lead-times, making them dependent on sufficiently stable and relatively high power prices. An increase in gas prices would increase the profitability of such investments, but Russia's gas price reform will probably have to follow a firm and credible development path before it can prompt large-scale shifts in investment incentives in nuclear and hydropower.

Demand Reactions Increasing as Reform Progresses?

On the other hand, it may be that reaction to demand has been limited because prices started changing from such low levels. Since the gas price has been kept so artificially low, gas has remained the cheapest alternative even *after* price increases, thus limiting any reduction in demand. This could indicate that current price movements cover a concave section of the Russian aggregate demand curve – in which case, price elasticity could be expected to increase with further price increases. However, the low current cost of gas and hence the relatively low share of gas expenditures in disposable income spending, together with the high competitiveness of gas compared to other fuels, would seem to indicate a relatively inelastic demand curve, which in turn would mean that gas consumption is likely to respond only moderately to an increase in prices. Moreover, significant action, accompanied by gas price reform, is needed in order to incentivize energy-efficiency measures that could shift the curve and thus create export potential. Lastly, the most important variable affecting gas demand is most likely GDP growth. Demand would probably lessen compared to what it *could* have been without gas price reform – but, with the Russian economy set to grow and a relatively inelastic demand curve, there is reason to question whether domestic gas demand will contract, reform or no reform.

Greenfield Developments

Replacement and expansion of total production is a second source of potential increased supply to European markets. Seen in isolation, price reform and increased prices should incentivize upstream investments, thereby increasing supply. However, investment decisions are not straightforward in Russia. From the supply side it does not automatically follow that investments in new production depend on domestic price levels – especially if they remain low, for whatever reason. The decision to invest in volumes for export depends on European prices, making domestic price reform irrelevant.

Background

In assessing whether the domestic price reform will be able to influence field developments it is essential to distinguish between Gazprom on the one hand and the other producers, the 'independents', on the other. The Russian domestic gas market is a *de facto* monopoly, with Gazprom the most important player by far. Remaining production is provided by companies known as 'independents', like Novatek, Itera and Nortgas, and oil companies that control some gas-producing assets or produce associated gas, such as Rosneft, Lukoil and TNK–BP.

The government regulates the gas market by requiring Gazprom to supply a certain amount of gas at regulated prices, thereby restricting the company's monopoly power vis-à-vis Russian consumers. Independent producers, by contrast, are not subject to price limits and are believed to have substantial potential for increasing their supply. However, the pipeline network is controlled by Gazprom – and that limits the market access of the independents.

Although Gazprom is required by the Federal Law on Gas Supply to grant access to the Unified Gas Supply System (UGSS) to all independent gas producers, this obligation is subject to the availability of UGSS capacity, as well as to compliance of the gas being transported with established quality and technical specifications, and the availability of connecting points to consumers. Since Gazprom controls the information on pipeline capacity, pipeline access to independent producers can be assumed to be granted at Gazprom's discretion. This in turn may be seen as an integral element in the company's optimization problem.

According to Gazprom, the volumes the independents transported through Gazprom-controlled pipelines decreased from 111 to 60 bcm from 2008 to 2009, rising slightly to 64.5 bcm in 2010 (Gazprom, 2012). Gazprom sales to the domestic market, however, contracted only 24.4 bcm from 2008 to 2009, even though market conditions should be the same for Gazprom and for the independents. To some extent, these disproportionate decreases in supply may be explained by the fact that Gazprom is obliged to supply at low regulated prices and that demand contractions that year did not support the higher prices expected by independents, reducing their supply. The degree to which independents can realize prices above average is believed to be limited, however: due to limited access to the pipeline network, they are led to sell gas to Gazprom at the wellhead at prices close to regulated levels. Or it may be that Gazprom deliberately increased its share of supply to the domestic market, limiting the independents' access, in order to compensate for revenue losses following reduced European demand.

Prices and Investment Dynamics

For the independents, the dynamics between domestic prices and production is a straightforward matter. Since Gazprom has a monopoly on exports, the independents can sell gas only on the domestic market, and the sole price of concern is the domestic price. Hence, a domestic price increase serves to incentivize their field development investments. For the independent producers, pipeline access is a far more pertinent issue than price reform. In fact, they are probably already limiting production due to restricted pipeline access, whereas they can expand production even without price reform – although higher domestic prices could lead these companies to invest more.

For Gazprom, the situation is more complex. Today Gazprom is required to supply a given volume at a given price, both decided by the authorities. However, if it is in mutual interest, Gazprom can let independent producers supply some of this volume. Whether Gazprom is willing to do so depends on the demand in export markets and Gazprom's total supply. It is rational for Gazprom to optimize its exports allocation first. Then, having done that, Gazprom can see what the existing production capabilities are, and allocate the remaining volumes to the domestic market. This gives Gazprom a clearly privileged position. If Gazprom had been constrained by the obligation to supply the domestic market, it could have bought gas from the independents or from central Asian countries if the cost was lower than marginal revenues in the European market. Independents are willing to supply the domestic market at prices lower than in the European market, proving that the cost of buying gas from these companies is relatively low.

Short-term Dynamics   In order to assess the impact of domestic prices on Gazprom investment decisions, it is necessary to consider several potential situations.

First we assume that domestic prices increase, but to a level lower than marginal revenue in the European market.[4] In that situation, both Gazprom and the independents would increase their revenues through higher windfall profits. Total supply, however, would not change, since Gazprom's supply is likely to be more costly and the company has already optimized its exports to a level where marginal costs equal marginal revenues. In other words, since the European prices do not change, there would be no signal to Gazprom to alter its volume allocation to Europe. Domestic market shares would remain unaltered, for the same reason.

Second, consider a domestic price increase that exceeds marginal revenue in the European market. Now Gazprom would have an incentive to increase its total supply until marginal costs equal the domestic price. To achieve this, it could

---

4   Marginal revenue is the increase in income of the sale of one more unit taking into consideration its price effect on all other units sold: that is, offering more units to the market creates a downward pressure on the price of all the units. Marginal cost is the increase in cost of producing the last unit.

restrict pipeline access for the independents and reduce their market share. Total supply would be the same, unless there were a reduction in demand.

Third, a domestic price increase might even imply less supply for the European market. If the domestic price rises to a level above marginal revenue in the European market and Gazprom's supply curve is dependent on one large project rather than a smooth curve, the marginal costs for Gazprom could become higher than both domestic and foreign prices. In that case, Gazprom would have an incentive to redirect volumes from Europe *to* Russia, thereby decreasing exports.

Fourth, Gazprom could hold back its total production after having optimized its exports while allowing the independents a share of domestic production and supply. This could be the result of an inter-temporal optimization decision to store volumes and supply the European market later. A domestic price increase could make the inter-temporal swap less attractive if the price increase were of sufficient magnitude. Nevertheless, also this would result in changes in domestic market shares rather than in increased total supply.

Long-term Dynamics   In the long-term perspective, the domestic price reform will probably have an impact. As production from current Russian gas fields declines, there will be a need for additional production from new fields to replace this dwindling production. If Gazprom were to boost domestic production through an increased market share this would not negatively affect the reserves of the independents, with the obvious exception of flaring.[5] The independents heed only the domestic price; and, as Gazprom will find itself forced to allow increasing volumes from the independents as its own production declines, the domestic price reform would help somewhat in propping up Russian long-term gas supply while also securing exports.

One main argument behind the gas price reform was that the domestic price was too low to help producers cope with future declines in production. Even though raising the price of gas will not in itself boost production, it will inevitably mean higher profits for Gazprom, making it increasingly possible for Gazprom to fund projects that are not necessarily economical from the company's point of view. To the extent that Gazprom is required to undertake non-profitable investments, increased prices could be necessary to secure adequate funding.

However, caution is required in assessing the potential impact of domestic gas price reform on investments in new deposits. Without pipeline access, an improved investment climate – and Russia ranks only 123 of 183 countries in the World Bank group's 'Ease of Doing Business' survey in 2010 (IFC and World Bank, 2011) – and a fiscal framework that supports investments, the development of new fields may be stalled even in a situation with equal profitability in the domestic and the foreign market.

---

5   Flaring gas is a byproduct of oil production and cannot be decreased unless oil production is decreased.

Possible Effects of Netback Pricing on Volume Choices in the European Market

The Federal Tariff Service publishes *would-be* reformed gas prices calculated by a special formula. This formula assumes that the domestic price will eventually depend on the European price, export taxation, transportation costs, customs duties and marketing expenses. The most influential variable in this equation is the 'average realized price in the European market'. Since the price to be determined by this formula does not reflect supply and demand in Russia, but rather European market conditions, there are implications for how Gazprom may act, insofar it can operate as a price-setter in the European market.

Market Power

Gazprom has since 2000 on average supplied some 20 per cent of EU consumption and 35 per cent of EU imports. The share of imports has been steadily decreasing from around 40 per cent in 2000 to 31 per cent in 2008. However, the volumes supplied have remained relatively unchanged. Moreover, as indigenous European production is set to decline, there will be a need for more Russian imports. The high market share gives grounds to expect substantial market power for Gazprom in the European market. Most economic analysis of Gazprom–EU relations assumes that other suppliers: indigenous, Norway, Algeria and others, are price-takers and supply as much as they can. However, what these producers can supply is not sufficient and Gazprom faces residual European demand over which the company has some market power; an increase in volumes will bring about a reduction in price, and vice versa. This assumption is supported by the fact that European demand for Russian gas decreased in the wake of the financial crisis. The supply from other exporters, such as Norway, did not contract similarly, thus highlighting Russia's role as the swing producer in the European market.

A stylistic framework is provided in Figure 7.5 and Figure 7.6 to analyse effects of Gazprom's allocation of volumes between domestic and European markets. Since domestic price and quantity is determined by the authorities and not by market conditions, Gazprom can only adapt to whatever is decided. The other gas producers then adapt to whatever Gazprom decides to do. For simplicity, it is assumed that the independents and other producers supply a fixed quantity of gas, starting in the left corner of Figure 7.5. Gazprom, with its export monopoly and *de facto* control over the domestic market, can optimize its allocation of gas between foreign and domestic markets. Q_T determines total quantity of Russian gas such that $Q_T - Q_D$ equals export potential. Now, consider the initial situation where domestic supply is at Q_{D1} and the regulated domestic price is set at P_{D1}. Then, the government decides that the domestic price is to be increased to P_{D2}, in preparation for netback pricing which is accompanied by a drop in domestic consumption from Q_{D1} to Q_{D2}. This would thus create a potential of $Q_{D1} - Q_{D2}$ that could be added to the exported volumes. However, increased supply would, *ceteris paribus*, imply a drop in foreign prices from P_{E1} to P_{E2} through an expansionary supply shock, given that the domestic demand reduction is completely channelled to European markets. With reference to the discussion above, this may not be an optimal choice for Gazprom.

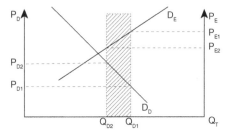

Figure 7.5   Gazprom's allocation of volumes

Consider the illustration of Gazprom's exports allocation depicted in Figure 7.6 where Gazprom initially faces a situation where marginal revenue (MR) equals marginal cost (MC_1) providing optimal volume Q_1 and price P_1. Now as a function of gas price reform, domestic demand contracts. These saved units may be considered 'free' on the margin, shifting marginal cost of exports from MC_1 to MC_2 and increasing supply from Q_1 to Q_2.

However, there is not necessarily a relationship between increased quantity of exports $Q_2 - Q_1$ and the no longer demanded domestic volume of $Q_{D1} - Q_{D2}$ from Figure 7.5. Thus, there should be no reason to expect that increased export supply will fully transform into increased exports. Considering Figure 7.6, it is clear that Gazprom would never

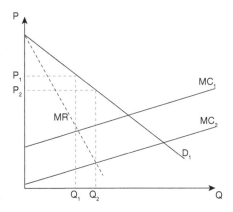

Figure 7.6   Monopolistic behaviour by Gazprom implies lower volumes to Europe

allocate supply to the right of the intersection of MR and MC. As long as domestic demand contraction is larger than the supply determined by this intersection, there will be excess exports supply. It follows that the more elastic the European demand, the larger will be the supply increase following a reduction in marginal cost. The extent of the increased exports is dependent on Gazprom's market power, the slope of European demand curve and the slope of Russian marginal cost curve. For example, with a relatively inelastic European demand curve, increased Russian supply would have a more limited impact on Russian exports.

However, the story does not end here. From a strategic perspective, since volume increases would most likely be sold on the spot market, they would contribute to strengthening the tendency of separation of oil and gas prices. Russia is interested in maintaining this link since the oil-linked contract prices received are relatively favourable. Thus, if the lower price of P_2 in Figure 7.6 contributes to breaking this link, the Gazprom allocation problem would imply less export than what is predicted.

Gazprom's optimization problem grows more complex with the implementation of the gas price reform. Since the netback tariff formula is tied to the average realized price on the European market, an increase of supply to Europe with a corresponding price decrease would inevitably lead to lower netback prices also on the domestic market. This implies the potential for a double revenue dip: not only will increased Russian supply create a downwards pressure on the European spot market, but it may also lead to lower revenues on the Russian market through the linkage of the netback price to European prices. In this situation, Gazprom may choose to supply less to the European market than would be the case when the European market is considered in isolation. An increase in European revenues would have to exceed the loss in Russian revenue from selling all domestic volumes at a lower price. As long as Russian demand is relatively inelastic, Gazprom will be ensured large domestic sales even at high prices which would increase the negative effect of a price decrease – whereas if demand were more elastic negative, revenue effects from price decreases would be partly compensated by increased sales. In a situation of steep Russian demand, Gazprom would have incentives to prop up European prices at the expense of lost export revenues, in order to tap into the domestic potential.

European Supply and Demand   Gas prices in Europe and the USA fell substantially in the wake of the financial crisis. More importantly, gas prices have failed to keep up with the oil price recovery, and the link between oil and gas prices has been weakened. The World Energy Outlook 2010 projected that the gas glut would reach 200 bcm in 2011 before it started slowly receding (IEA, 2010). These predictions were modified in the next year's edition: the gas glut would dissipate sooner than expected in 2011, thanks to a sharp rebound in gas demand (IEA, 2011). Nevertheless, the rising supply of unconventional gas and LNG, with Europe as its major outlet, puts pressure on conventional gas suppliers (IEA, 2012). This expected decrease in market power would translate into more price-elastic European demand with respect to Russian volumes, and thus a greater potential export-volume effect of contractions in domestic demand. Concessions made by Gazprom and other producers to cut prices and ease take-or-pay obligations may indicate that Gazprom has been trying to maintain its market share and adjust to new market realities.

In line with the argumentation in Chapters 3 and 6, in the short run price differences will exist in different gas markets in the world. However, in the long run, with LNG markets becoming a more integrated part of the global gas market, there will be a tendency towards price equalization throughout the world. Thus, the export prices that Russia can obtain in Europe will be dependent not only on European demand, but also on global gas demand and supply. With the ensuing gas glut and projections of limited increase in demand, there is reason to remain cautious about the prospects for gas price increases in Russia's export markets. This is connected not only to temporal contraction of demand caused by the recent economic and financial crisis, but to new technological developments enabling increased production of non-conventional gas in the USA and possibly elsewhere (the EU, China, Latin America), and with globalization of the LNG market (Grivach, 2012).

What about Political Risk?

Russia has been repeatedly accused of using the dependency of importing countries on Russian gas exports as a tool of its foreign policy, making Europe focus more on security of supply and the need for diversification (Larsson, 2010; Stegen, 2011; Perovic, Orttung and Wenger, 2009). Most cited are the 'gas wars' with Ukraine, which were followed by fears of repeated disruptions of supply to Europe, and the reliability of Russia was questioned. With dependency on Russia appearing less and less attractive, the EU started to intensify its efforts to look into other sources of energy. However, this view fails to take adequate account of the dependency of Russia and Gazprom on the European market for revenues. In fact, it may be argued that in a situation with low domestic gas prices, Russia has become more dependent, as exports to Europe were the only source of revenue sufficient to keep the Russian gas machine running (Goldthau, 2008; Orttung and Overland, 2011).

Some statistics can help in getting an overview over how important Russia is for Europe, and vice versa.

Figure 7.7 shows Russian exports by country in 2011. We see that the largest importer of Russian gas that year was Germany, followed by Ukraine. The only other EU country that imports over 10 bcm per year is Italy. Other major importers were Turkey and Belarus. Combined, these seven top importers took roughly 62 per cent of Russian exports in 2011.

Not surprisingly, countries geographically close to Russia are the most dependent on Russian gas as a percentage of total gas imports. Several EU members are heavily dependent on Russian gas to meet domestic demand; moreover, of the European countries that import all their gas from Russia, only Ukraine and Belarus are outside the EU. In 2009, 13 countries got more than half their gas imports from Russia; both Germany and Italy had shares above 30 per cent.

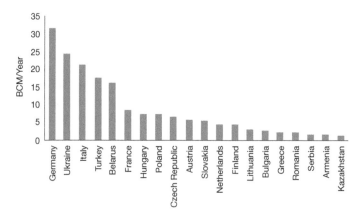

Figure 7.7   Russian gas exports in bcm, by country (2011) (BP Statistical Yearbook, 2012)

However, a different picture emerges when we look at the share of Russian gas as a percentage of total energy consumption. Belarus and Ukraine are most dependent on Russian gas; other East European countries such as Hungary, Romania and Slovakia also have a high share of Russian gas in their total energy consumption. By contrast, turning to major revenue-generating countries like Germany, Italy and France, we see that Russian gas is responsible for less than 10 per cent of their total energy consumption; and in the whole EU, Russian gas imports cover only between 5 and 6 per cent of total energy consumption. This means that Russia may be far more dependent on Western European demand than the Western European countries are dependent on Russian supply.

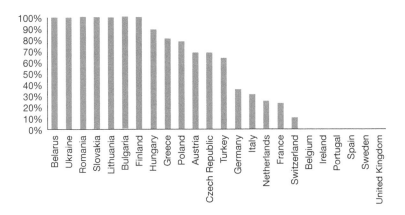

Figure 7.8   Russia's share of total gas imports (BP 2010, authors' calculations)

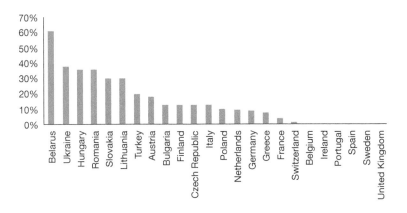

Figure 7.9   Dependence on Russian gas as % of total energy consumption, selected countries (BP, 2010 and authors' calculations)

While there is little doubt that Russia, at least to some extent, uses its gas exports as a political tool in its dealings with neighbouring countries, there can be even less doubt that Russian gas exports have been vital in securing revenues for Gazprom and for the Russian state. In the period 2000–2009, roughly two-thirds of Gazprom's revenues were generated by sales to Europe (excluding former Soviet countries) (Gazprom, 2011). Moreover, the profit margin has most likely been higher for these volumes than those sold domestically. There is therefore reason to expect Moscow to have treated seriously EU fears about Russia's ability to provide long-term supply and subsequent plans of diversifying away from Russian gas. As pointed out by Morbee and Proost (2009), increased unreliability of Russian gas supply hurts Russia through both reduced contract volumes *and* reduced prices; moreover, Europe is hurt as well, due to reduced security of supply.

The 21 April 2010 agreement with Ukraine on linking gas rebate with the issue of the Black Sea Fleet bases on the Crimean Peninsula could therefore be interpreted as a double success: it cemented Russia's position as the leading power among the former Soviet countries, while simultaneously allaying European fears of supply disruptions. The result was paradoxical – by using gas in a clearly political manner, the Russian government to some extent lessened the political risk it attached to Russian exports arising from its repeating quarrels with Ukraine. However, as shown by developments in the second half of 2011 and since, not all problems in Russian–Ukrainian gas relations have been settled. On the one hand, President Yanukovich and his team have proven less willing to accommodate Moscow's interests; on the other hand, Russia has managed to get the Nord Stream built, the first strategic pipeline to give direct access to its most important energy customer, Germany, making it less dependent on transit countries.

Domestic price reform may have surprising implications for those concerned with Russia's politicized approach to gas exports. Thus far, Gazprom and Russia have had to rely on exports to Europe in order to subsidize domestic consumption – but gas price reform could increase Gazprom's domestic revenues and reduce its dependence on export revenue. With Europe becoming increasingly dependent on Russian gas and Russia becoming less dependent on exports to Europe, the relative strength in this gas relationship may shift in Russia's favour. European support for launching the gas price reform in Russia as a remedy for European security of gas supply may thus create an unwanted side-effect.

Conclusions

The Russian gas universe is complex, and no single factor is likely to alter the future of Russia's supply, demand and export capabilities. Although the future of gas price reform itself is uncertain, three main arguments indicate at best a limited increase in exports even if the reform is implemented – and some factors could indicate that exports might contract.

There is reason to expect limited income effects of demand, as many consumers have no possibilities for curbing their own consumption. Today most Russians pay a utility fee that is independent of the volume of gas consumed. Moreover, many households can simply not adjust their gas consumption: supply is determined either for the building or even the area, and the only way to regulate indoor temperature may be to open the windows. Substitution effects hold perhaps the greatest potential for increasing the reserves base. Energy-efficiency measures have a potential of curbing demand by 45 per cent. However, many investments that should generate attractive returns are simply not being made. Moreover, the slow and erratic pace of the gas price reform fails to provide incentives to energy-efficiency efforts, as it creates severe timing issues for industries that might be contemplating investments. Fuel switching could reduce the demand for Russian gas – but the alternatives are not necessarily viable options. Coal creates local pollution through lower air quality and ash disposal; moreover, to be profitable, the deposits must be located relatively close to demand centres. Nuclear and hydropower are alternatives, but long lead-times and uncertain reform progress limit the impact of gas price reform on investment decisions here. All in all, evidence concerning gas price elasticity is scarce in Russia. Gas consumption, fuelled by GDP growth, has actually increased simultaneously with gas prices. Thus, while gas price increases might reduce demand compared to what it *could* have been, there is perhaps less reason to expect any absolute reductions in demand.

Second, increased domestic gas prices will, at least in the short term, not serve to incentivize investments in new production. The only impact of the price reform will be windfall profits to Gazprom and independent companies, as well as possibly shifting the shares of domestic production towards Gazprom. If the domestic price rises to a level above marginal revenue in Europe, Gazprom may even cut back on exports in favour of a higher share of the domestic market. For the other producers, price reform will increase the willingness to invest in new production – but, without access to pipelines, these actors cannot boost their production and sales. Nevertheless, in the long term the production potential of Russia could increase, as the independents could develop more resources than they would have without gas price increases. Eventually, these reserves will be utilized, as Gazprom's assets will start to decline and the independents' market shares will increase.

Third, the most influential variable concerning Russian exports is foreign prices. Russia has possibilities for influencing the prices received from Europe since it currently functions as a swing producer. Increased supply would most likely be directed to the spot market, thus putting further pressure on the gas price. Weakened spot prices would place greater strains on the already weakening link between oil and gas prices that Russia is keen to maintain to avoid pressure on its oil-indexed contracts. Furthermore, if the gas price reform is implemented in its current form, lower European prices would inevitably result in lower domestic prices as well, thereby creating a double revenue dip.

Lastly, Russian gas price reform could even act to curb exports. As the domestic market becomes as profitable as foreign markets, Gazprom's domestic profits will increase. This could make Gazprom, and Russia, less dependent on sales on foreign markets. Russia has been accused of using its dominant position as a gas supplier as a political tool. For example, it has allegedly penalized disobedient countries with higher gas prices in times of turbulent bilateral political relations, as in the case of Ukraine after the Orange Revolution. Thus far, however, Russia has had to employ this political tool with caution, precisely because most of Gazprom's profits have been generated abroad. With the share of profits generated in foreign markets diminishing, Russia would gain an improved bargaining position towards its foreign customers. Thus, domestic gas price increases may come with an unexpected, and in foreign eyes unwanted, side-effect: Russia could become more inclined to utilize gas as a political weapon in its foreign relations.

References

Ahrend, R. and W. Tompson 2004. Russia's gas sector: the endless wait for reform? OECD Working Paper No. 402. Paris: OECD.

Andresen N.A. 2008. Public choice theory, semi-authoritarian regimes and energy prices: a preliminary report, RUSSCASP Working Paper [online] Oslo: Fridtjof Nansen Institute FNI. Available at http://www.fni.no/russcasp/WP-2008-010_NAA_Public_choice_theory.pdf [accessed 15 June 2012].

Bashmakov I. 2005. Energeticheskaya effektivnost' v Rossii i perspektivi eksporta rossiyskogo gaza [Energy effectiveness in Russian and prospects of exporting of Russian gas] [online: Center for Effective use of energy]. Available at: http://www.cenef.ru/file/GasExportsProspects.pdf [accessed 15 June 2012].

BP 2010. *BP Statistical Review of World Energy* [online: BP]. Available at: www.bp.com [accessed 15 June 2012].

Christie, E.H. 2010. The Russian gas price reform and its impact on Russian gas consumption. Pan-European Institute Electronic Publications. Turku: Pan-European Institute [online]. Available at: www.tse.fi/pei.

Eurostat 2011. Energy production and imports. Available at: http://epp.eurostat.ec.europa.eu/statistics_explained/index.php/Energy_production_and_imports.

Fjaertoft, D.B. 2010. Gas reform and the industrial lobby: gas price dependency of the Russian steel industry. RUSSCASP Working Paper. Oslo: Fridtjof Nansen Institute.

Gazprom 2010. Gazprom_databook_-rus_1h_2010 [online: BP]. Available at: www.gazprom.com [accessed 15 June 2012].

Gazprom 2011. *Gazprom in Figures 2006–2010 Factbook*. Moscow: Gazprom.

Gazprom 2012. Gazprom's questions. Available at: http://www.gazpromquestions.ru/fileadmin/files/2011/view_version_06.02.2012.pdf [accessed 15 June 2012].

Goldthau, A. 2008. Resurgent Russia? Rethinking energy Inc. Policy Review [online] 147, pp. 53–63. Available at: http://www.hoover.org/publications/policy-review/article/5714 [accessed 15 June 2012].

Grivach, A. 2012. Direktor Platts Jorge Montepeque: Vy ne uznayete gazovoy rynok cherez pyat' let [Director of Platts, Jorge Montepeque: 'you will not recognize the

gas market in five years']. Moskovskie Novosti, 25 January. Available at: http://
mn.ru/business_oilgas/20120125/310366907.html [accessed 15 June 2012].

Henderson, J. 2011. *Domestic Gas Prices in Russia – Towards Export Netback?*
NG 57. Oxford: Oxford Institute for Energy Studies.

IEA 2010. *World Energy Outlook 2010*. Paris: International Energy Agency.

IEA 2011. *World Energy Outlook 2011*. Paris: International Energy Agency.

IEA 2012. *World Energy Outlook 2012*. Paris: International Energy Agency.

IFC and World Bank 2011. Doing business. Available at: http://www.doingbusiness.
org/rankings [accessed 15 July 2012].

Kutschera, H. and I. Øverland 2009. Pricing pain: prospects for reducing
subsidies for natural gas in Russia, RussCasp Working Paper. Oslo: Fridtjof
Nansen Institute.

Larsson, R. 2010. *Rysk Energimakt – Korruption og säkerhetsfixering i nationens
namn*. Stockholm: Ersatz.

Lelushkin, N. 2011. Zaprogrammirovannaya bessistemnost'. Rynok sam po
sebe ne mozhet reshit' problemu energoeffektivnosti vysshego poryadka
[Programmed unfunctionality. Market will not solve the problems of energy
effectiveness by itself]. *Nezavisimaya Gazeta NG–Energia*.

Mitrova, T. 2009. Natural gas in transition: systemic reform issues, in Simon
Pirani (ed.) *Russian and CIS gas markets and their impact on Europe*. Oxford:
Oxford University Press.

Morbee, I. and S. Proost 2009. Russian gas imports in Europe: how does Gazprom
reliability change the game? CES Discussion Paper, 15 July. Available at:
https://lirias.kuleuven.be/bitstream/123456789/160888/6/DPS0802newvs.pdf
[accessed 29 June 2012].

Orttung, R.W. and I. Overland 2011. A limited toolbox: explaining the constraints
on Russia's foreign energy policy. *Journal of Eurasian Studies*, 2(1), pp. 74–85.

Oxenstierna, S. 2010. *Russia's Nuclear Energy Expansion: User Report*.
Stockholm: Totalförsvarets forskningsinstitut FOI.

Perovic, J., R.W. Orttung and A. Wenger (eds) 2009. *Russian Energy Power
and Foreign Relations: Implications for Conflict and Cooperation*. London:
Routledge.

RusHydro (n.d.) Hydropower in Russia. Available at: http://www.eng.rushydro.ru/
industry/history [accessed 29 June 2012].

Sagen, E.L. and M. Tsygankova 2006. Russian natural gas exports to Europe.
Discussion Paper No. 445. Oslo: Statistics Norway.

Sagen, E.L. and M. Tsygankova 2008. Russian natural gas exports – will Russian
gas price reforms improve the European security of supply? *Energy Policy*,
36(2), 867–80.

Sargsyan, G. and Y. Gorbatenko 2008. Energy efficiency in Russia: untapped
reserves. World Bank Group with CENEF. Available at: http://www.ifc.org/
ifcext/rsefp.nsf/AttachmentsByTitle/FINAL_EE_report_Engl.pdf/$FILE/
Final_EE_report_engl.pdf [accessed 29 June 2012].

Stegen, K.S. 2011. Deconstructing the energy weapon: Russia's threat to Europe
as case study. *Energy Policy*, 39, 6505–13.

Chapter 8

The Future of Russian Gas Production: Some Scenarios

Eini Laaksonen, Hanna Mäkinen and Kari Liuhto

Introduction

Natural gas is a major energy resource for the European Union, currently accounting for about one quarter of its primary energy consumption. Furthermore, this share is likely to stay approximately the same during the next 20 years (European Commission, 2010). However, the EU's own gas production is declining, even when gas from EEA-member Norway is included. As a result, the gap between supply and demand for natural gas may grow, bringing greater dependence on imported energy. Indeed, the EU is already rather heavily dependent on Russian energy: as the holder of the world's largest gas reserves, Russia currently accounts for slightly more than one-third of the EU's gas and oil imports, and more than 30 per cent of its coal imports (Eurostat, 2011).

Thus, Russia has a dominant position in the European energy market and both Russian energy production and EU energy supply are tightly interlinked. However, Russia is beginning to face new, serious challenges regarding its gas industry – the West Siberian basin is reaching its peak in production, and the resource bases in northern and eastern territories require investment and time before they can produce enough gas to offset the gradual decline of the old basins. Moreover, the global economic situation has started to improve and energy demand is rising again – not only in Europe, but, for instance, in China as well. In addition, domestic demand for natural gas continues to increase in Russia.

Russia intends to invest somewhere between USD 1,800 and 2,100 billion in the oil and gas industry by 2030 to increase production in order to keep up with projected European and Asian demand and with its growing domestic consumption (*Eurasia Daily Monitor*, 2009). In fact, Gazprom hopes that its share of European gas market will reach 30 per cent by 2020 (RIA, 2012), as against 27 per cent in 2011 (Gazprom, 2012d). Gazprom also expects eventually to reach a similar market share in Asia (*Moscow Times*, 2010). During the economic crisis, gas deliveries to Europe fell drastically but have since been increasing again and are approaching the pre-crisis level of 185 billion cubic meters (bcm). In 2011, Gazprom supplied a total of 150 bcm of gas to Europe, which was 8 per cent more than in 2010. The largest buyers of Russian gas were Germany, Turkey and Italy (Gazprom, 2010b; 2012b). For instance, between 2001 and 2009 Russia's share of

EU gas imports declined from 48 per cent to 34 per cent, whereas Norway's share rose from 23 per cent to 31 per cent (Eurostat, 2011).

The supply issue is complicated by the fact that any planned increases in gas production and the required investments are always linked to projected foreign sales. The recent boom in unconventional gas resources has created a revolution in US gas production, with the resultant uncertainties in the global gas business creating new conditions for the development of new gas fields in Russia. In the pre-crisis year of 2008 Gazprom produced approximately 550 bcm of natural gas whereas in 2009 the production sank to record lows of 461 bcm (Pettersen, 2010). Gazprom has cut its gas production forecasts several times during the past few years and now expects its gas production to regain pre-crisis levels only in 2014 (Gazprom, 2012e). To meet the growing demand for natural gas, Russia will have to step up the gas production of Gazprom and the independent producers, increase its gas imports from Central Asia, and improve domestic energy efficiency – all challenging and time-consuming tasks. It seems uncertain whether Russia will be able to meet the rising energy demands of European and Asian markets in the near future.

This chapter begins by identifying the main factors affecting the future of Russia's role as dominant energy supplier to the EU, and then offers scenarios for the future of Russian gas production.

The Future of Russian Gas Production

Supply Outlook of Existing Gas Fields

The main engine of Russian gas production is the Nadym-Pur-Taz (NPT) region in Western Siberia, which accounts for approximately 90 per cent of all gas production in Russia. Production in the region is dominated by the fields Urengoy, Yamburg and Medvezhye. These have been central to Russian gas production for two decades, but production is starting to decline. To offset the decrease in the short term, new and smaller fields are being commissioned in the area, the most productive being Zapolyarnoye and Yuzhno Russkoye fields (Söderbergh, Jakobsson and Aleklett, 2010). However, new major fields will need to be brought into production in Russia within the next ten years if the country is to meet the demand for natural gas.

Currently the future of Russian gas production rests in the Yamal gas fields, which form the core of the country's future production capacity. If the development of these resources faced significant delays, the effects would soon become apparent in Russia's export capacity for gas as feeding the domestic markets is the top priority of the state (Putin, 2012). As estimated by Söderbergh, Jakobsson and Aleklett (2010), even a five-year delay from the announced start date of 2012 would by 2017 result in a gap of some 100 bcm in the Russian annual natural gas supply capacity. As well as the Yamal fields, most of the potential new natural gas deposits are located far from centres of consumption and in environments where operations are more difficult than in the fields of western Siberia. Consequently, production costs are likely to be higher in the future, in turn entailing a greater need for investment. Foreign know-how and finance are required for the development of the Russian gas industry, particularly in terms of challenging offshore fields. Will there be enough new gas production in Russia for the next ten years and beyond?

Yamal Gas Fields Proceed

The natural gas fields of the Yamal Peninsula are crucial for the future of Russian gas production: they hold 16,000 bcm of explored gas reserves with condensate reserves estimated at 230.7 million tons. Gas from this region would be enough to satisfy world demand for five years. According to Gazprom estimates, Yamal will be producing 360 bcm of natural gas annually by 2030, and its reserves will suffice for 50 years (Gazprom, 2009a). Gas from the Yamal fields is to be piped for domestic consumption in Northwest Russia and partly exported to Europe. Originally the liquefied natural gas (LNG) production from the peninsula was planned to be shipped to Asia, Europe and the USA, although at the moment the USA is becoming increasingly self-sufficient in terms of gas supply. Vladimir Putin as well as Gazprom CEO Alexei Miller have stated that the Yamal project intends to target East Asian gas markets as the first priority because these markets have the greatest commercial potential (Socor, 2009a).

The largest single field in the Yamal Peninsula is the Gazprom-owned Bovanenkovo, located 40 km from the coast of the Kara Sea and holding 4,900 bcm of gas (Gazprom, 2009b). Natural gas production from this field is estimated to reach 115–140 bcm annually. Realization of this megaproject commenced in 2007 (Gazprom, 2009c). In addition to Gazprom, Russia's number two gas producer, Novatek, controls significant gas reserves on the Yamal Peninsula. Unlike Gazprom, Novatek is a private company, but it enjoys good political connections. While Gazprom proceeds with the Bovanenkovo field on the west side of the peninsula, Novatek is planning production at its Yuzhno-Tambeyskoe field on the eastern side. However, Novatek has been restricted in delivering gas abroad since Gazprom controls Russian exports of gas. In 2010, the two companies agreed on constructing an LNG plant on Yamal and Gazprom Export acting as an agent for export sales (LNG World News, 2010; Novatek, 2012). Despite the agreement, Novatek has been actively working to get its own export rights for Russian gas (Shiryaevskaya and Clark, 2012).

Nevertheless, foreign technology and finance are needed for the construction of gas liquefaction plant, LNG port, and a fleet of tankers and icebreakers. With the help of President Putin, both Gazprom and Novatek have been seeking assistance in terms of investment and technology from several foreign companies. Russia seems to be proposing a deal – access to the Yamal resources, in return for technology transfer, asset swaps and cross-investment possibilities (Socor, 2009b). Such deals have been made in the past as well, as European energy companies could otherwise have had difficulties in gaining access to Russian oil and gas fields. The Russian side has been particularly interested in companies involved in energy logistics, as in this way Russian energy corporations can be able to control strategic parts of the energy supply (Liuhto, 2010).

In 2011, French Total joined Novatek's Yamal LNG project with a 20 per cent share. As long as it retains a controlling stake, Novatek seems willing to invite also other partners into the project (Total, 2011). The project includes the construction of production, storage and loading facilities for LNG, based on the resources of the South Tambeyskoye field. The launch of the plant's first production line is scheduled for 2016 (Novatek, 2011). In addition to this field, Total and Novatek are jointly developing the Termokarstovoye gas field in the Yamal Peninsula, with the expected start-up in 2015 (Total, 2011).

In addition to the uncertain market conditions, a difficult issue is the major infrastructure projects that need to be carried out in parallel with development of the gas fields. For instance, in order to deliver gas from Gazprom's Bovanenkovo field, it will be necessary to build a 2,450 km gas transmission system, including the Bovanenkovo-Ukhta gas transmission corridor linking the field with the Russian pipeline grid and further with Europe via the Yamal-Europe pipeline and Nord Stream. Particularly given the remoteness and lack of adequate infrastructure, the Yamal fields are more challenging, expensive and time-consuming to develop than were Russia's existing fields. According to some estimates, full development of the region might require investments of up to USD 180 billion by the year 2035, with Bovanenkovo alone needing 12 billion and the transport system 90 billion. The financial side seems somewhat problematic for the main operator Gazprom, which finds itself faced with significant investment expectations (development of the Yamal fields, the Shtokman field, fields in East Siberia and the Far East), and the construction of export pipelines to Europe and China (Terletski, 2010; Staalesen, 2012c).

In 2009, Gazprom decided to postpone the commissioning of the Bovanenkovo field and the Bovanenkovo-Ukhta gas trunkline system from 2011 to 2012 due to the uncertain forecasts of future consumption of natural gas (Gazprom, 2009c). Both Gazprom and Novatek have requested tax concessions from the government to enable them to meet all their state-initiated production obligations, and, in 2010, the Russian Finance Ministry proposed zero tax regimes for 12 years for gas extraction and liquefaction activities in the Yamal-Nenets district (Guha, 2010). In 2011, Dmitry Medvedev signed into law various oil and gas tax breaks, including scrapping the extraction tax for LNG production in the Yamal Peninsula (Soldatkin, 2011). Nevertheless, production of Yamal gas is expected to prove significantly more expensive than the current gas from Nadym-Pur-Taz region (Terletski, 2010).

Even though project postponements seem to be caused by uncertainties in the global gas market, it is clear that Yamal gas will be needed soon, as production from the NPT region is decreasing. According to Gazprom (2012a), the construction of new gas transmission facilities as well as preparations to launch gas production in Bovanenkovo have now proceeded successfully. The welding, installation and testing operations have been finished at the first string of the Bovanenkovo–Ukhta pipeline, and the opening ceremony took place in October 2012. However, to reach the expected volumes of 115 bcm by 2017, Russia needs to step up the field's development (*The Siberian Times*, 2012).

Shtokman: Uncertain Future

The other major potential source of gas in Russia is the Shtokman gas field project in the Barents Sea. According to Shtokman Development AG (2013), Shtokman reserves account for 3,800 bcm of gas and about 37 million tonnes of gas condensate – enough to meet total EU gas demand for over seven years. Annual production from the project is envisioned to reach 70 bcm of natural gas and 0.6 million metric tons of gas condensate, which is only slightly less than the entire gas output of Norway. However, the field is located some 600 km northeast of the city of Murmansk, in extreme Arctic conditions under the Barents Sea, at depths varying from 320 to 340 meters. The gas is to be piped 550 km to the shore and further for local use in Northwest Russia, to Europe via the Nord Stream pipeline, and to be turned into LNG at a plant in Teriberka near Murmansk. Originally, the LNG was to be shipped to North America (Shtokman, 2009), but with the 'shale gas revolution' other markets will have to be found.

The license to explore and produce gas and gas condensate in the Shtokman field is owned by Gazprom. However, Shtokman Development AG, which is to develop the Shtokman field and to be the owner of its first-phase infrastructure for 25 years after its commissioning, is controlled by an international consortium. It was registered in 2008 with Gazprom owning 51 per cent, French Total 25 per cent, and Norwegian StatoilHydro 24 per cent of the company stock. The total costs of the project are expected to reach USD 30 billion, with USD 15 billion required already in the first phase (RIA, 2009). Due to the expected high costs and the market-related uncertainties, the project has faced several postponements and the companies involved have pressed for a special tax regime to be put in place (Amos, 2011).

Furthermore, the consortium partners have been reported as disagreeing on several key technical matters. An important issue has been a legal solution regarding the inclusion of foreign companies in the project. On joining the project, Total and Statoil were not awarded shares in the reserves itself, but shares in the company that would develop the field. Shtokman Development is to develop and operate the field and then own the built infrastructure during the project's first phase (for 25 years after production has started); thereafter, everything will be handed over to Gazprom (Laaksonen, 2010). According to the reported plans, half of project costs are to be paid during the first phase, whereas the peak production of 71 bcm per year is expected to be reached only after 25 years (Shtokman, 2009). All that means that the project setup does not seem overly favourable to its foreign partners.

Indeed, in July 2012, Statoil transferred its 24 per cent of Shtokman Development AG shares to Gazprom, thereby leaving the project, in accordance with the expiry of the initial shareholder agreement (Shtokman, 2012). Statoil's decision did not come as a surprise, since the Shtokman project had appeared troublesome with years of delays. Moreover, Statoil has more promising fields to develop, for instance in the Norwegian areas of the Barents Sea. Statoil had earlier expressed its doubts over Shtokman cost estimates, development models and economic viability without tax concessions (Nordic Intel, 2012). Moreover, the company wanted Shtokman to produce only LNG instead of pipeline deliveries (Staalesen, 2012e). At the same time, Gazprom had been hinting at the possibility of inviting other partners, particularly Shell, into the consortium by reducing the shares of the current foreign partners (Staalesen, 2012d). Gazprom and Shell have cooperation experience from building together Russia's first LNG plant in the Sakhalin II offshore gas project (Nordic Intel, 2012).

Although Gazprom recently indicated intentions as to 100 per cent LNG production (Staalesen, 2012d) and to having the consortium involved in the development of all the three phases of the project, the Norwegian partner is now out of Shtokman Development (Pettersen, 2012). While Total also has other major projects to focus on in the Russian Arctic, including the Yamal LNG with Novatek (Staalesen, 2012b), it has not returned its Shtokman shares, and has been undertaking further technical studies aimed at enabling an economically viable project (Regan, 2012). Altogether the three partners have spent an estimated USD 1.5 billion on project development (Staalesen, 2012a).

The uncertain situation on the global gas market and the wobbly project consortium have resulted in speculations on the future of Shtokman. In August 2012, Gazprom stated that developing the Shtokman field was uneconomic for the time being (Kramer, 2012) and that the project's phase one had been put on hold. The investment decision has been postponed indefinitely (Macalister, 2012). However, it is possible that in the future, as global gas supplies start to decrease and as new technologies emerge, developing the expensive but huge Shtokman field may again become relevant.

Is Yamal Enough?

Considerable uncertainty still attends the future development of gas demand and prices (see e.g. MIT, 2011). Is it reasonable to begin production in the costly Shtokman field in the foreseeable future? Should the development of Yamal fields be further stepped up?

It is clear that the giant fields that are the core of Russia's gas production are declining and will have to be replaced. In addition to European consumption, Russian domestic demand for natural gas continues its slow increase, despite the significant potential for energy-saving measures. According to UN estimates, Russia has the potential to reduce its annual domestic natural gas consumption by 240 bcm, which would far exceed the volume of the country's annual gas exports in 2005–2008 (UNDP, 2010). Although the Russian leadership has acknowledged the importance of energy saving and has incorporated it into the country's energy strategy, effective implementation of energy-saving measures and imbedding energy-saving attitudes in the Russian mentality will take years, if not decades (Mäkinen, 2010a). Furthermore, as long as the domestic price of gas remains low, improving energy efficiency is not a matter of great interest for Russian industrial and household consumers.

Development of the Yamal fields, which has been prioritized over Shtokman, is far more crucial for Russia's future as a gas producer. If development here does not proceed as planned, that will have serious impacts on the economy and on Russia in general – the state depends on hydrocarbons for two-thirds of its exports, half of its federal budget, and 20 per cent of its GDP. Pumping oil and gas is the only way to maintain balance in the Russian political system (Hulbert, 2012). Work on the Bovanenkovo field has been proceeding as planned, and there is no room for further delays. Already by 2020, over 200 bcm annually will have to be produced from the new fields in order to offset the decline in Russia's depleting basins. If this level is not reached by that time, Russia may face problems in meeting its own demand for gas as well as its international contractual commitments.

The progress of Yamal and the decisions related to Shtokman will have a huge impact on Russia's position as a reliable, long-term strategic energy partner. An important input in those decisions is a clearer understanding of probable future developments on the regional and global gas markets.

Future External Demand for Russian Gas

The European Union: Still Important

The economic crisis reduced the external demand for Russian gas in 2009 and created a short-term oversupply of piped gas in the European gas market, further reinforced by the increased inflow of LNG earlier intended for the US market. Significant volumes of spot gas became available on the European gas market, which depressed spot prices (Eurogas, 2010). Natural gas consumption in the EU decreased by some 6 per cent in 2009 compared to 2008 – from 490 bcm in 2008 to 460 bcm in 2009. In 2010, gas consumption grew to pre-crisis levels of 497 bcm, but was down by almost 10 per cent in 2011 – to 448 bcm (BP, 2012). According to Eurogas (2010), it will take several years for demand to regain the high levels recorded a few years earlier. Therefore, as a result of both the economic crisis and the growing emphasis on renewable energy and energy efficiency, long-term natural gas demand in the EU is expected to be 15–20 per cent lower than forecast in 2007.

Even so, natural gas is expected to remain an important energy source for the European Union. The share of natural gas in total EU energy consumption is likely to remain relatively stable, around a quarter for the period 2010 to 2030.

At the same time, EU gas production is expected to decrease from 182 bcm in 2010 to 123 bcm in 2020 and further to 84 bcm by 2030 (European Commission, 2010). Figures provided by Eurogas differ slightly, indicating an even more rapid decline for natural gas production in the EU: from 196 bcm in 2007 to 113 bcm in 2020 and further to 63 bcm by 2030 (Eurogas, 2010). Indeed, by 2030 the EU is expected to require 40 per cent more gas imports than in 2005. Still, the impact of the economic decline is evident – in 2007 import demand had been estimated to increase by 70 per cent. Recent lower estimates are due to the decreased economic activity and consumption during the economic crisis which reduced energy consumption, the accelerated energy efficiency progress during and after the economic recovery, and new carbon emissions reduction policies adopted by the EU (European Commission, 2010).

Future gas supplies from Norway, the EU's second largest gas supplier, also give cause for worry. With domestic supply in the EU declining and the EU attempting to reduce its energy dependence on Russia, demand for Norwegian gas is expected to grow. However, estimates as to the future development of Norwegian gas production vary – with a peak somewhere between 2015 and 2020. As the majority of the large fields currently in production are now starting to deplete, further development of Norwegian gas production will depend on the discovery of new major fields and the pace of investments (Söderbergh, Jakobsson and Aleklett, 2009). Although several new offshore gas fields have been discovered in Norway in recent years (see e.g. Staalesen, 2010; Reuters, 2012; WSJ, 2012), the rate of new discoveries will need to improve and new fields must be put into production in order to replace today's depleting fields. Thus, the potential for increasing gas imports from Norway to the EU would seem limited (Söderbergh, Jakobsson and Aleklett, 2009).

All this indicates that by 2030 some 70 per cent of natural gas consumed in the EU will have to be imported from outside Europe (Eurogas, 2010). The EU can face difficulties in meeting its gas demand in the future, and may have to search for additional gas supplies (Table 8.1).

The future of European gas demand will also depend on the development of the EU's energy mix and climate policies. Germany's decision to abandon nuclear power by 2022 can increase the demand for natural gas, since the energy gap

Table 8.1   EU-27 natural gas supply outlook 2007–2030, bcm

	2007	2015	2020	2025	2030
Additional supplies to be defined *(environmental scenario)*	0	0	28	32	39
Additional supplies to be defined *(base case)*	0	0	14	60	105
Contracted imports and prolongations from outside Europe	210	266	300	304	289
Available volumes from Norway	79	94	108	103	98
EU indigenous production	196	155	113	84	63

Source: Eurogas (2010) and authors' calculations.

caused by the nuclear shutdown will have to be closed somehow. Curbing climate change requires reductions in carbon dioxide emissions and a shift from fossil fuels towards renewable sources of energy. However, because natural gas is the cleanest of the fossil fuels, it has been widely advocated as a bridge between coal/oil and renewables. Indeed, the EU Energy Roadmap 2050 accords to natural gas a key role in the transformation of energy system – emissions can be reduced with existing technologies by substituting coal and oil with natural gas until at least 2030 or 2035 (European Commission, 2011). In the short and medium term, the share of natural gas in the EU's energy structure may well rise, although in the long term the decarbonisation of the energy system could lead to reductions in natural gas consumption as well. Thus, for the time being, Russia seems likely to remain an important supplier of gas to Europe.

Asia: Rising Demand for Natural Gas

From the Russian perspective, another significant phenomenon is the potential emergence of new buyers for its gas. China has emerged as an important player in the development and use of natural gas resources, and is increasingly interested in Russian gas supplies. At today's levels of production and consumption, China's own natural gas reserves – 3,100 bcm – are set to last for almost 30 years. However, natural gas consumption in China has increased over tenfold during the past ten years, reaching 130 bcm in 2011 (BP, 2012), and is projected to exceed 310 bcm by 2035. China's imports of natural gas have increased dramatically in recent years, jumping from a 12 per cent share of consumption in 2010 to 22 per cent in 2011 (EIA, 2012b).

Consequently, along with oil, Russia is about to start exporting natural gas to China from its fields in Eastern Siberia and the Russian Far East. Since 2007, Gazprom has been executing a state-run Eastern Gas Program, aimed at developing an integrated gas production, transportation and supply network in Eastern Siberia and the Russian Far East, where gas transport system has been rather limited compared to the western regions of Russia. The Eastern Gas Program also takes account of potential gas exports to the Asia–Pacific region (See Söderbergh, Jakobsson and Aleklett, 2010; Gazprom, 2012c). Since 2006, Russia and China have engaged in negotiations on supplies of natural gas, but have as yet failed to agree on the pricing for gas. As an alternative to pipeline supplies, Russia started to negotiate on LNG supplies to China in the beginning of 2013. LNG might be supplied by Gazprom, which is building an LNG plant close to Vladivostok, as well as from the Yamal LNG facility, co-owned by Novatek and Total (*Moscow Times*, 2013).

On the other hand, Japan is also interested in Russian gas, particularly LNG, which it is already importing from Russia. After the 2011 Fukushima nuclear accident, the demand for fossil fuels has increased in Japan and, while negotiations between Russia and China have stalled over the price of gas, Japan may well increase its gas imports from Russia (Golubkova and Pinchuk, 2012).

At present, the plan seems to be for Russia to apportion its declining supply of natural gas among its domestic markets, the EU and Asian economies. Central Asia has been an important source of inexpensive natural gas for Russia, but has recently begun to reassert its role as a major production area of natural gas. Russia, the EU and China are all actively engaged in competition for Central Asia's natural gas to meet their growing demands. China has already begun gas imports via the pipeline from Turkmenistan and has agreed on gas supplies from Kazakhstan and Uzbekistan, strengthening its position in the area (Henderson, 2011).

The United States: Growth in Natural Gas Production

The most significant phenomenon in recent developments in the global gas sector has been the exploitation of unconventional sources of natural gas, which has seen rapid growth in the USA since 2000 due to rapidly expanding production of shale gas. In 2005 unconventional gas – shale gas, tight gas and coalbed methane – accounted for nearly half of total US natural gas production and in 2010 already almost 60 per cent (IEA, 2012). This dynamically increasing production of unconventional gas has now made USA the world's biggest gas producer, surpassing Russia in 2009, 2010 and 2011 (BP, 2012).

Shale gas in particular has attracted considerable attention. Shale gas resources in North America are huge and production from shale formations is the fastest-growing source of unconventional natural gas production, whereas relatively little change in production is expected from other unconventional sources. In 2010, the USA produced 141 bcm of shale gas, which accounted for 23 per cent of its total dry gas production. The US Energy Information Administration (EIA, 2012a) estimates that shale gas will make up 49 per cent of total US dry gas production by 2035, totalling 385 bcm. However, the technological uncertainties surrounding the production may either speed up or slow down the projected growth in shale gas.

Abundant unconventional gas resources have already brought deep changes in the US energy market, such as increased self-sufficiency in natural gas and lower natural gas prices. Only a few years ago the USA was thought to be growing increasingly dependent on imported LNG, so LNG suppliers around the world began building up their production and import capacities (Brown, 2010). If natural gas production continues to grow as projected, US natural gas production will exceed consumption around the year 2020 (EIA, 2012a). With the rise in gas production, import and export prices were more than 50 per cent lower in 2009 compared to 2008, and have stayed approximately at the same level in 2010 and 2011 (EIA, 2012c). Replacing other energy sources by 'home-grown' natural gas can indeed enhance US energy security (Brown, 2010).

Moreover, there can be implications for climate policy. At a power plant, combustion of natural gas produces half as much CO_2 emissions as burning coal. By replacing coal with gas in its power production, the USA can cut its CO_2 emissions significantly, by up to 20 per cent, at relatively low cost (MIT, 2011).

On the other hand, many uncertainties surround the production of unconventional gas. The lifespan of shale gas wells might be shorter than with conventional gas wells – evidence here is still contradictory, and further technological developments are likely to bring improved possibilities for exploitation of reserves (Stevens, 2010). In the USA, the environmental impacts of drilling for shale gas have already given rise to concern, and local opposition to production has emerged. For example, citing the risks to groundwater, New York State has placed a temporary moratorium on a technology used in shale gas drilling – horizontal hydraulic fracturing. This technique requires the use of chemicals and vast amounts of water to release natural gas from rock (Navarro, 2010), and may have negative impacts on the environment, the groundwater in particular. Environmental regulations may well be tightened in the future, becoming a major obstacle to the further expansion of shale gas production in the USA and elsewhere.

The surge in US natural gas production is likely to have profound implications for the global natural gas market. Because the USA is able to reduce its gas imports, the LNG exporters of the Middle East and Asia must find other markets for their LNG supplies in Europe and Asia. If the unconventional gas production boom in the USA continues, the country even has the potential to start exporting natural gas to Europe and Asia in the future. This could be motivated by the substantial price difference of gas in the European and Asian markets compared to US markets (Liuhto, 2012). Recently-built LNG import terminals for re-gasification have been standing almost empty in the USA, and their owners are considering the possibilities of converting them into export terminals. If the USA increased its LNG imports, that could add volatility to a global gas market already undergoing significant transformations (McNulty, 2010). These recent developments have put pressure on the pricing policy of Gazprom, forcing it to add spot gas prices to its long-term contracts and to agree to see up to 15 per cent of its gas sales to Europe at spot prices (Forest, 2010). Oversupply of gas and the gap between spot prices and long-term contracts also threaten investments in new fields and pipelines in Russia. Gazprom has already postponed some projects, including the development of the Shtokman field.

Possible Impacts of Unconventional Gas Production in the EU

The shale gas revolution in the USA has given rise to debate on how far this might be replicated elsewhere. Almost half of the world's recoverable natural gas resources are estimated to be unconventional, mainly in the form of shale gas. Unlike conventional gas, found mainly in Russia and the Middle East, unconventional gas resources are believed to be widespread across the globe (*Economist*, 2012). If exploitable, they could change the global energy scene profoundly. For example, according to one recent assessment, the former USSR area is believed to hold shale gas resources even larger than those of the USA (World Energy Council, 2010). As an illustration of the worldwide division of shale gas resources, one assessment is presented in Table 8.2 (based on World Energy Council, 2010).

Table 8.2   Estimated shale gas resource potential, 2010 (1 tcm [trillion m³] equals 1,000 bcm)

Region	Estimated shale gas resources potential, tcm
North America	126.6
Latin America	10.6
Western Europe	15.8
Central and Eastern Europe	15.8
Former Soviet Union	153
Central Asia and China	10.5
Pacific	21.1
Middle East and North Africa	37.0
Sub-Saharan Africa	28.8

Source: World Energy Council (2010).

Because of the promising resource assessments, interest in shale gas has been growing, especially in Europe and in China. For instance, the estimated 15,000 bcm of shale gas in Western Europe would be enough to supply the EU27 for about 30 years at current consumption rates. However, very little is as yet known about shale gas reserves outside the USA, and shale gas has not yet been produced commercially in other areas. The EIA (2011) has already presented varying estimates on shale gas resources worldwide which have generally been more conservative, and the figures may still change significantly once more thorough-going assessments have been conducted.

In the EU shale gas exploration is underway in Austria, Germany, Hungary, Poland, Sweden and United Kingdom, among others. European energy companies, among them Statoil and Total, have also started joint ventures with their US counterparts to acquire know-how on shale gas production. One of the most active EU countries regarding shale gas has been Poland, assessed (EIA, 2011) to hold around 5,300 bcm of technically recoverable shale gas resources. At current consumption rates, that would be enough to supply Poland for over 300 years. However, the recent assessment of the Polish Geological Institute (2012) on the probable range of recoverable shale gas looks significantly less promising – between 346 and 768 bcm. Political and media hype has been significant in Poland, and major energy companies like Chevron, ConocoPhillips and ExxonMobil have been exploring there. However, ExxonMobil recently ended its shale gas tests in Poland, not finding enough shale gas for commercial production (Cienski, 2012). For Poland these less optimistic assessments can be a setback. The country is today heavily dependent on Russia for its gas supplies, and success in shale gas production could increase its self-sufficiency significantly, thereby enhancing its energy security.

There are many barriers and challenges to be dealt with before shale gas production can take off in the EU. As population density in the EU is much higher than in the USA, disruptions caused by shale gas drilling are likely to give rise to local opposition, especially because of environmental concerns; moreover, environmental legislation in the EU is far stricter than that of the USA. It seems certain that debate on the environmental effects of shale gas production will be even more heated in the EU (Kefferpütz, 2010). Indeed, France, Romania and Bulgaria have already instituted environmentally-motived moratoriums on shale gas exploration (Cienski, 2012). The geology of much of Europe also differs from that of North America, which could make it much less promising for shale gas production, and as yet there is not much specific geological data or experience concerning shale gas. The EU also lacks the equipment and technology, as well as the skilled workforce of the USA. This may prove a significant constraint, as having to bring in expertise from abroad can slow down the development of shale gas. Moreover, high costs of labour and drilling can make production less economically attractive. And finally, as yet there have been no significant political incentives to promote the development of shale gas in the EU (Kefferpütz, 2010). Huge investments are needed across the globe for shale gas exploration and exploitation, and the legal issues in particular need to be resolved in order to attract investments (World Energy Council, 2012).

Gazprom CEO Miller has publicly stated that the company does not believe that shale gas production will have a decisive impact on the European gas market or challenge the predominance of pipeline gas (RIA, 2010). Gazprom has emphasized the negative aspects involved, such as environmental impacts, rapid depletion rates and high production costs. All the same, Gazprom has acknowledged the need to follow developments in the shale gas industry worldwide (Gazprom, 2010a). China is believed to hold large shale gas resources and is interested in developing them to help in meeting its growing energy demand. China has worked together with US experts on hydraulic fracturing practices and invited several Western companies to take part in shale gas development. The first shale gas wells have already been drilled in China. According to its latest Five-Year Plan, China aims to produce 6.5 bcm of shale gas by 2015, and increase production at least tenfold by 2020 (Yang, 2012). The development of unconventional gas in China and other Asian countries could undermine Gazprom's plans for supplying large volumes of Russian gas to the Asian market.

As yet, however, unconventional gas does not seem to threaten Gazprom's position as a significant supplier of natural gas. Whatever the size and recoverability of shale gas reserves worldwide, it will take a long time before any significant production can take place outside the USA (Mäkinen, 2010b). We have already noted some major obstacles in the EU, particularly as regards environmental concerns. Moreover, even if shale gas production may prove possible in the EU, the volumes will not be enough to counter the decline in conventional gas production in the short term. Thus the EU will continue to rely on pipeline imports, possibly from Russia and Norway, and on LNG. On the other hand, although shale gas may not prove to be a game-changer for the whole EU, it might have a greater impact on the energy situation of individual countries, such as Poland. Due to the major uncertainties surrounding shale gas production, its future developments are hard to predict. With appropriate new technology, unconventional gas might become 'conventional' in the future – and could threaten Gazprom's market position in Europe and worldwide.

Conclusions

Russian gas production and EU demand for natural gas are tightly interlinked. Figure 8.1 summarizes the main issues affecting both demand and production.

The future of the EU–Russian gas relationship is strongly tied to the prospects for production of unconventional gas resources that could have significant effects on the EU's energy import policy. However, the many questions surrounding the availability of unconventional gas as well as the economic and environmental feasibility of such projects bring significant uncertainties for the whole gas market.

A further important factor is the potential of other energy suppliers to the EU. For instance, gas supplies from the USA would be welcome, as would new additional volumes from Norway, Central Asia or the Middle East. However, Russia's well-established share of EU energy imports and the fact that Russia has the greatest gas reserves in the world is likely to give it an important role also in the next decade. Still, Russian pricing policies and interest in basing gas trade on fixed long-term contracts might lead the EU to look for alternative suppliers as well as alternative sources of energy. In addition, of course, the EU's future climate policies will affect its choice of energy sources in the years to come.

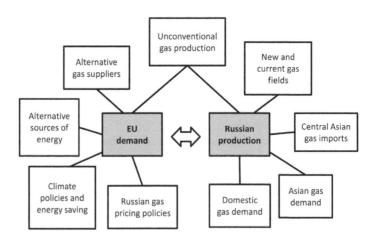

Figure 8.1   Russia as the future supplier of natural gas to the EU

To Russia, the boom in unconventional gas is not a welcome development. Due to the accompanying uncertainties, Russia has shown some reluctance to develop its new and more expensive gas fields – but these fields would become crucially important if the development of the EU's shale gas projects were to prove problematic. Another question is whether Russia will continue to balance supplies between East and West. Might it turn increasingly towards China, Japan and other growing East Asian economies? The Northern Sea Route has potential for becoming a gas transit corridor from the Russian Arctic gas fields to the Asian markets. The highest bidder will also get access to the Central Asian gas resources, which, through Russian deliveries, are a significant source of the EU's energy supply. These energy commodities are also vital to Russia for covering its own gas needs and meeting its international commitments. Moreover, despite the significant potential for energy-saving measures in Russia, domestic demand for gas does not seem to be in decline: indeed, it is expected to grow.

As Russia will not be able to bring its challenging Arctic gas fields into production on its own, the country's policy towards foreign investors and international energy cooperation will be central to the future development of the energy sector. Russia has welcomed foreign investors in its huge gas projects, but still maintains control over the gas resources, requiring its foreign partners to share their expertise and provide the necessary financial resources. Consequently, the returns on the foreign companies' investments remain uncertain, particularly since foreign companies in Sakhalin I, Sakhalin II and Kovykta have encountered serious problems with the state and Gazprom. On the other hand, European energy companies do not have many alternative production areas, and seem increasingly interested in participating in Russian projects. Russia gains in bargaining power when it has many potential partners to choose from. Russia also seems to aim for long-term agreements with the foreign, at least partly state-owned, gas companies participating in its projects. In this way, Russia's partners hope to be able to spread the greatest risks entailed in these gas projects – price instability and sales uncertainties on the international gas market (Socor, 2009a).

The future of Russian gas production is linked to the development of its new major gas fields. Here we may note three options:

1. *Yamal completed as scheduled, Shtokman cancelled.* According to current estimates, gas produced in Yamal could meet demand for the next decades, providing that the Yamal projects are completed on schedule. However, if the US demand for gas imports were to rise, there might not be enough gas for the EU, China and the USA. Gas would then go to the highest bidder – or to the most important political ally.

2. *Yamal completed according to schedule, and Shtokman development begins.* Completing projects on both Yamal and Shtokman fields will require extensive investments and intensive international cooperation, and Shtokman is currently not viable as a project. If the production of unconventional gas remains profitable in the USA and extends to the EU,

China and Russia as well, developing these basins would not be profitable. Producing too much gas would result in lower gas prices, causing serious harm to the Russian economy.

3. *Yamal development encounters significant delays.* If the development of the Yamal fields cannot meet the schedule, for technical or financial reasons, Russia will face serious problems. Yamal will have to be in full production already by 2020 to cover domestic consumption and provide exports to replace falling production in Russia's other fields. Domestic gas consumption in Russia continues to increase, and improved energy efficiency is unlikely to have significant effects here in the next ten years or so. Moreover, if Shtokman is not to be developed in the near future, Russia will not be able to afford delays in the development of the Yamal fields.

We see that there are two extreme scenarios for the future: too much gas in the market, or not enough. In the first case, where the supply of gas is higher than demand, the price of natural gas will fall drastically. That would have serious consequences for the Russian economy, due to the high cost of developing its new gas fields and lower returns on that investment. Gas exports, and even more so oil exports, generate a significant share of Russia's budgetary revenues; if energy commodity prices were to drop, the country might face a financial crisis that could also shake its governing bodies.

The second extreme scenario, in which the supply of natural gas cannot meet market demands, would not be any better option. A shortfall of only 200 bcm in the supply chain would cause trouble, as the Russian domestic market would consume almost all the gas produced, with nothing left for exports. Such a supply shortfall could be caused, for instance, by a five-year delay in the development of the Yamal gas fields. Gas exports are and will remain crucial to the Russian economy. If Yamal development is delayed, Russia may face lack of gas for export, thus lowering its export revenues. The consequences might be similar to the case above: in addition, new problems would emerge with Russia's international relations, particularly with the EU and with China.

This means that Russia will have to get the new fields producing gas in time, while also ensuring against oversupply. Balancing those two interests will not be easy, given all the uncertainties related to the future of global gas production. However, Russia must do its best to find ways of getting optimal amounts of gas produced. By 'optimal' is meant amounts that would be enough to satisfy the domestic as well as foreign markets, at profitable prices. This would guarantee a stable future for Russia – but the country would remain heavily reliant on its energy sector, and that would undermine the drive towards diversification of the economy. Russian energy efficiency reform would not be in a hurry either, even though natural gas is undeniably a limited resource. Nevertheless, in the current situation, if the optimal balance between supply and demand cannot be found, the consequences are likely to be even worse. Table 8.3 summarizes the three scenarios.

Table 8.3 Future Russian gas production: three scenarios

	1. GAS REVOLUTION	2. GAS EVOLUTION	3. GAS STAGNATION
CAUSES	Yamal fields proceed according to schedule; also Shtokman is developed Shale gas booms not only in North America, but on other continents as well Due to the increased flow of LNG, the gas glut continues in the European market Demand for Russian gas decreases in Europe and Asia	Yamal developed in time, Shtokman in the future 'Optimal' amounts of gas produced in Russia, i. e. supply almost meets demand	Yamal development delayed several years Shtokman is cancelled Gas glut on the European market a short-term phenomenon Shale gas not fulfilling optimistic expectations in Europe Demand for gas imports increases in the EU and Asia Russian domestic consumption continues to grow
CONSEQUENCES	Price of gas falls Gas sold increasingly at spot prices, not long-term contracts Export volumes are high, but the profits are not Extensive gas field investments do not pay back: Gazprom and Russian Federation face a financial crisis Foreign investors face significant losses with the gas projects	Russia can charge profitable prices for gas exports Russia able to supply gas both to East and West, as well as its domestic market Foreign investors receive return on their investments Russian public may not be forced to improve energy efficiency Russia not forced to diversify its economy	Gas prices skyrocket Russia might not have enough gas for exports, at least not enough to fulfil all its agreements Forced and rapid reduction of Russian gas consumption, perhaps bringing disaffection and societal problems Crisis for the Russian economy and for political stability Crisis for European energy security

All three are possible, although the odds cannot be estimated. Moreover, these scenarios can be seen as extreme development paths: reality is often not as black-and-white. Nevertheless, we see them as a useful way of highlighting and summarizing important issues related to the future of gas production in Russia. Also the EU should follow the situation closely, because if Russian attempts at starting new gas production fail, the EU might face an energy crisis. Production within the EU is in decline, and Norwegian production alone cannot counteract that, let alone compensate for the lack of gas supplies from Russia. Shale gas might provide a surprising solution, but there is no guarantee that this could happen soon. Shale gas is unlikely to reduce the EU's need for gas imports in the near future, in any case. Nor should the EU stand aloof: the future of both the EU and Russia – and their energy cooperation – is at stake.

References

Amos, H. 2011. Statoil presses for Shtokman decision. *The Moscow Times* [online], 7 November. Available at: http://www.themoscowtimes.com/business/article/statoil-presses-for-shtokman-decision/447185.html#ixzz1d0eUSQtx [accessed 29 June 2012].

BP 2012. Statistical review of world energy 2012. Available at: http://www.bp.com/sectionbodycopy.do?categoryId=7500&contentId=7068481 [accessed 29 June 2012].

Brown, S.P.A. 2010. Abundant natural gas could mean a paradigm shift in U.S. energy markets and policy, *Resources* [online] 175, summer. Available at: http://www.rff.org/Publications/Resources/Pages/default.aspx [accessed 15 June 2012].

Cienski, J. 2012. ExxonMobil ends shale gas tests in Poland. *Financial Times* [online], 18 June. Available at: http://www.ft.com/intl/cms/s/0/5e883fdc-b94c-11e1-b4d6-00144feabdc0.html#axzz294TyMIFc [accessed 15 June 2012].

Economist 2012. Focus: natural gas reserves [online] Available at: http://www.economist.com/blogs/graphicdetail/2012/06/focus [accessed 15 November 2012].

EIA [US Energy Information Administration] 2011. World shale gas resources: an initial assessment of 14 regions outside the United States [online]. Available at: http://www.eia.gov/analysis/studies/worldshalegas/pdf/fullreport.pdf [accessed 15 June 2012].

EIA [US Energy Information Administration] 2012a. *Annual Energy Outlook 2012 with Projections to 2035*. Available at: http://www.eia.gov/forecasts/aeo/pdf/0383%282012%29.pdf [accessed 15 June 2012].

EIA [US Energy Information Administration] 2012b.Country analysis: China. Available at: http://www.eia.gov/countries/cab.cfm?fips=CH [accessed 15 June 2012].

EIA [US Energy Information Administration] 2012c. U.S. natural gas prices. Available at: http://www.eia.gov/dnav/ng/ng_pri_sum_dcu_nus_a.htm [accessed 15 June 2012].

Eurasia Daily Monitor 2009. Russian oil and gas in 2030. 28 September, 9(164).

Eurogas 2010. Long term outlook for gas demand and supply 2007–2030. Available at: http://www.eurogas.org/figures_statistics.aspx [accessed 15 June 2012].

European Commission 2010. EU energy trends to 2030 – update 2009. Available at: http://ec.europa.eu/energy/observatory/trends_2030/index_en.htm

European Commission 2011. Energy roadmap 2050. Available at: http://ec.europa.eu/news/energy/111219_en.htm [accessed 15 June 2012].

Eurostat 2011. Energy production and imports. Available at: http://epp.eurostat.ec.europa.eu/statistics_explained/index.php/Energy_production_and_imports [accessed 15 June 2012].

Forest, D. 2010. Gazprom finally accepts that shale gas has changed the world, *Oilprice* [online], 3 March. Available at: http://oilprice.com/Energy/Gas-

Prices/Gazprom-Finally-Accepts-that-Shale-Gas-Has-Changed-the-World. html [accessed 15 June 2012].

Gazprom 2009a. Railroad significance for the Yamal project implementation. Available at: http://gazprom.com/production/projects/mega-yamal/railway/ [accessed 15 November 2012].

Gazprom 2009b. Yamal megaproject. Available at: http://www.gazprom.com/ production/projects/mega-yamal/ [accessed 15 June 2012].

Gazprom 2009c. Gazprom resolves to postpone Bovanenkovo field commissioning. Available at: http://www.gazprom.com/press/news/2009/june/article66599 [accessed 15 July 2012].

Gazprom 2010a. Gazprom to keep focus on shale gas industry development. Press release, 23 November. Available at: http://www.gazprom.com/press/ news/2010/november/article105829/ [accessed 15 July 2012].

Gazprom 2010b. Gazprom in questions and answers. Available at: http://eng. gazpromquestions.ru/index.php?id=4 [accessed 15 November 2012].

Gazprom 2012a. First string of Bovanenkovo – Ukhta linepipe welded up. http://www.gazprom.com/press/news/2012/may/article136037/ [accessed 20 July 2012].

Gazprom 2012b. Gazprom in foreign markets. Available at: http://eng. gazpromquestions.ru/index.php?id=4 [accessed 20 July 2012].

Gazprom 2012c. Gazprom keeps actively implementing eastern gas program. Press release, 13 December. Available at: http://www.gazprom.com/press/ news/2012/december/article151548/ [accessed 18 February 2013].

Gazprom 2012d. Gazprom to continue reinforcing its standing in Europe. Press release, 25 April. Available at: http://www.gazprom.com/press/news/2012/ april/article134112/ [accessed 20 July 2012].

Gazprom 2012e. Mineral and raw material base development. Gas production. Gas transmission system development. Press Conference, 23 May. Available at: http://www.gazprom.com/f/posts/76/977901/development-resources-stenogram-2012-05-23-en.pdf [accessed 15 November 2012].

Golubkova, K. and D. Pinchuk 2012. Russia, Japan back LNG; could delay export pipelines. *Reuters* [online], 8 September. Available at: http://www.reuters.com/ article/2012/09/08/us-russia-lng-asia-idUSBRE88705R20120908 [accessed 1 March 2013].

Guha, S. 2010. Russian government may offer tax breaks to Yamal LNG project. BSR Russia [online]. Available at: http://www.bsr-russia.com/en/oil-gas/ item/1345-russian-government-may-offer-tax-breaks-to-yamal-lng-project. html [accessed 15 June 2012].

Henderson, J. 2011. The pricing debate over Russian gas exports to China. *Oxford Institute for Energy Studies* [online]. Available at: http://www.oxfordenergy.org/ wpcms/wp-content/uploads/2011/10/NG-561.pdf [accessed 1 March 2013].

Hulbert, M. 2012. Arctic oil: Putin's last chance. *European Energy Review* [online], 19 July. Available at: http://www.europeanenergyreview.eu/site/ pagina.php?id=3813 [accessed 15 June 2012].

IEA 2012. Golden rules for a golden age of gas. World energy outlook special report on unconventional gas. Available at: http://www.worldenergyoutlook. org/media/weowebsite/2012/goldenrules/WEO2012_GoldenRulesReport.pdf [accessed 15 June 2012].

Kefferpütz, R. 2010. Shale fever: replicating the US gas revolution in the EU, *CEPS Policy Brief* [online], No.210/June 2010. Available at: http://www.ceps. eu/book/shale-fever-replicating-us-gas-revolution-eu [accessed 15 July 2012].

Kramer, A.E. 2012. At Gazprom, views conflict on viability of Barents Sea gas project. *The New York Times* [online]. Available at: http://www.nytimes. com/2012/08/31/business/global/gazprom-natural-gas-project-in-arctic-seas-in-doubt.html?_r=0 [accessed 15 June 2012].

Laaksonen, E. 2010. Political risks of foreign direct investment in the Russian gas industry – the Shtokman gas field project in the Arctic Ocean. Electronic Publications of the Pan-European Institute, 14/2010. Available at http://www. tse.fi/FI/yksikot/erillislaitokset/pei/Documents/Julkaisut/Laaksonen_netti_ final.pdf [accessed 15 July 2012].

Liuhto, K. 2010. Energy in Russia's foreign policy. Electronic Publications of Pan-European Institute [online], 10/2010. Available at: http://www.tse.fi/FI/ yksikot/erillislaitokset/pei/Documents/Julkaisut/Liuhto_final.pdf [accessed 15 June 2012].

Liuhto, K. 2012. A liquefied natural gas terminal boom in the Baltic Sea region? Electronic Publications of the Pan-European Institute [online], 5/2012. Available at: http://www.tse.fi/FI/yksikot/erillislaitokset/pei/Documents/ Julkaisut/Liuhto%20LNG.pdf [accessed 15 June 2012].

LNG World News 2010. Russia: $20 billion for Yamal LNG terminal. Available at: http://www.lngworldnews.com/russia-20-billion-for-yamal-lng-terminal/ [accessed 15 July 2012].

Macalister, T. 2012. Plug pulled on Russia's flagship Shtokman energy project. *The Guardian* [online]. Available at: http://www.guardian.co.uk/ world/2012/aug/29/shtokman-russia-arctic-gas-shale [accessed 15 July 2012].

MIT 2011. The future of natural gas. An interdisciplinary MIT study. Available at: http://mitei.mit.edu/publications/reports-studies/future-natural-gas [accessed 18 February 2013].

The Moscow Times 2010. Asia may use as much gas as Europe. 5 February.

The Moscow Times 2013. Kremlin in talks on LNG exports to China. 25 February. Available at: http://www.themoscowtimes.com/news/article/kremlin-in-talks-on-lng-exports-to-china/476061.html [accessed 1 March 2013].

Mäkinen, H. 2010a. Energy efficiency – a new energy resource for Russia? Electronic Publications of the Pan-European Institute [Online], 20/2010. Available at: http://www.tse.fi/FI/yksikot/erillislaitokset/pei/Documents/Julkaisut/Makinen_ Energy%20efficiency%20in%20Russia.pdf [accessed 15 June 2012].

Mäkinen, H. 2010b. The future of natural gas as the European Union's energy source – risks and possibilities. Electronic Publications of Pan-European

Institute [online], 9/2010. Available at: http://www.tse.fi/EN/units/specialunits/pei/publications/Pages/2010.aspx [accessed 15 July 2012].

McNulty, S. 2010. US to take on rivals in natural gas, *Financial Times*, 7 October. Available at: http://www.ft.com/cms/s/0/6be97200-d22f-11df-8fbe-00144feabdc0.html#axzz1BbYo1iW1 [accessed 15 July 2012].

Navarro, M. 2010. N.Y. Assembly approves fracking moratorium, *The New York Times*, 30 November. Available at: http://green.blogs.nytimes.com/2010/11/30/n-y-assembly-approves-fracking-moratorium/#more-82409 [accessed 15 June 2012].

Nordic Intel 2012. Why did Norway's Statoil abandon the Shtokman natural gas project? Available at: http://nordicintel.com/statoil-abandons-shtokman-projec/ [accessed 15 July 2012].

Novatek 2011. NOVATEK closes sale of interest in Yamal LNG to TOTAL. Available at: http://www.novatek.ru/en/press/releases/index.php?id_4=433 [accessed 15 August 2012].

Novatek 2012. South-Tambeyskoye field [online: Novatek]. Available at: http://www.novatek.ru/en/business/yamal/southtambey/ [accessed 15 August 2012].

Pettersen, T. 2010. Gazprom cuts gas production forecast, *Barents Observer* [online], 22 December. Available at: http://barentsobserver.com/gazprom-cuts-gas-production-forecast.4864957-16178.html [accessed 15 July 2012].

Pettersen, T. 2012. Total: Shtokman decision postponed until 2013. *Barents Observer* [online], 4 May 2. Available at: http://barentsobserver.com/en/energy/total-shtokman-decision-postponed-until-2013 [accessed 15 July 2012].

Polish Geological Institute 2012. Assessment of shale gas and shale oil resources of the Lower Paleozoic Baltic-Podlasie-Lublin Basin in Poland. *First report*. Available at: http://www.pgi.gov.pl/pl/dokumenty-in/cat_view/294-aktualnoci-2012/297-zasoby-gazu.html [accessed 15 August 2012].

Putin, V. 2012. Prime Minister Vladimir Putin meets with Deputy Prime Minister Igor Sechin and deputy chairmen of Gazprom Management Committee Alexander Medvedev and Andrei Kruglov. Transcript of the beginning of the meeting on the Official Website of the Russian Federation. Available at: http://government.ru/eng/docs/17990/ [accessed 15 November 2012].

Regan, J. 2012. Total says Shtokman project not delayed indefinitely. *Reuters*. Available at: http://www.reuters.com/article/2012/08/31/total-shtokman-idUSWEA112220120831 [accessed 15 July 2012].

Reuters 2012. Production starts at Norway's Atla gas field. 8 October. Available at: http://uk.reuters.com/article/2012/10/08/norway-atla-idUKL6E8L8EV220121008 [accessed 15 July 2012].

RIA 2009. Business investment in Russia's Shtokman gas field hits $500 mln. *RIA Novosti*. Available at: http://en.rian.ru/business/20090922/156215557.html. [accessed 15 May 2012].

RIA 2010. Shale gas no threat to Gazprom in Europe – CEO, *RIA Novosti*. Available at: http://en.rian.ru/russia/20100409/158500290.html [accessed 15 May 2012].

RIA 2012. Gazprom's exports to Europe come at a higher price in 2012. *RIA Novosti*. Available at: http://rt.com/business/news/gazprom-europe-export-plan-582/ [accessed 15 May 2012].

Shiryaevskaya, A. and T. Clark 2012. Billionaire Timchenko says Novatek pushing for export rights. *Bloomberg* [Online]. Available at: http://www.bloomberg.com/news/2012-05-15/billionaire-timchenko-says-novatek-pushing-for-export-rights-1-.html [accessed 15 June 2012].

Shtokman 2009. *Shtokman Development AG*. Available at: http://www.shtokman.ru/eng/about [accessed 15 May 2012].

Shtokman 2012. *Statoil ASA transferred its shares of 'Shtokman Development AG' to Gazprom*. Available at: http://www.shtokman.ru/en/press/news/2012/242/ [accessed 15 June 2012].

Shtokman Development AG. 2013. *Shtokman gas and condensate field*. Available at: http://www.shtokman.ru/en/project/gasfield/ [accessed 1 March 2013].

The Siberian Times 2012. *Giant new Siberian gas field starts production to supply Europe*. Available at: http://siberiantimes.com/business/profile/news/giant-new-siberian-gas-field-starts-production-to-supply-europe/ [accessed 4 March 2013].

Socor, V. 2009a. Putin touts vast Yamal Project for global export of LNG. *Eurasia Daily Monitor*, 6(176).

Socor, V. 2009b. Putin's Yamal offer: a preliminary assessment. *Eurasia Daily Monitor*, 6(177).

Soldatkin, V. 2011. Russia's President approves oil and gas tax breaks. *Reuters*. Available at: http://www.reuters.com/article/2011/07/21/russia-energy-tax-idUSLDE76K14E20110721 [accessed 15 May 2012].

Staalesen, A. 2010. New gas field smaller than expected. *Barents Observer* [Online]. Available at: http://barentsobserver.com/en/sections/energy/new-gas-field-smaller-expected [accessed 15 May 2012].

Staalesen, A. 2012a. Gazprom shakes up Shtokman Development. *Barents Observer* [Online]. Available at: http://barentsobserver.com/en/energy/gazprom-shakes-shtokman-development [accessed 15 May 2012].

Staalesen, A. 2012b. Goodbye, Shtokman. *Barents Observer* [Online]. Available at: http://barentsobserver.com/en/energy/goodbye-shtokman-08-08 [accessed 15 May 2012].

Staalesen, A. 2012c. In Yamal, the world's fourth most expensive energy project. *Barents Observer* [Online]. Available at: http://barentsobserver.com/en/energy/2012/12/yamal-worlds-fourth-most-expensive-energy-project-03-12 [accessed 18 February 2013].

Staalesen, A. 2012d. New partners for Shtokman. *Barents Observer* [Online]. Available at: http://barentsobserver.com/en/energy/new-partners-shtokman [accessed 15 May 2012].

Staalesen, A. 2012e. We want only LNG at Shtokman, says Statoil. *Barents Observer* [Online]. Available at: http://barentsobserver.com/en/energy/we-want-only-lng-shtokman-says-statoil [accessed 15 July 2012].

Stevens, P. 2010. The 'shale gas revolution': hype and reality, Chatham House Report. Available at: http://www.chathamhouse.org.uk/publications/papers/view/-/id/947/ [accessed 15 May 2012].

Söderbergh, B., K. Jakobsson, and K. Aleklett 2009. European energy security: The future of Norwegian natural gas production, *Energy Policy*, 37(12), 5037–55.

Söderbergh, B., K. Jakobsson, and K. Aleklett 2010. European energy security: an analysis of future Russian gas production and exports, *Energy Policy*, 38(12), 7827–43.

Terletski, V. 2010. Gazprom's hands ever deeper in state's coffers. *RusBusiness News* [online]. Available at: http://www.rusbiznews.com/news/n744.html [accessed 15 May 2012].

Total 2011. Russia: Total and Novatek to join forces for development of Yamal LNG. Available at: http://www.total.com/en/about-total/news/news-940500.html&idActu=2656 [accessed 20 May 2012].

UNDP 2010. National human development report in the Russian Federation 2009: energy sector and sustainable development. Moscow: UNDP Russian Office.

World Energy Council 2010. Survey of energy resources: focus on shale gas. Available at: www.worldenergy.org/documents/shalegasreport.pdf [accessed 15 June 2012].

World Energy Council 2012. Survey of energy resources: shale gas – what's new. Available at: http://www.worldenergy.org/documents/shalegasupdatejan2012.pdf [accessed 15 May 2012].

WSJ (*Wall Street Journal*) 2012. Poland's PGNiG says new gas field found off Norway's shore. *Wall Street Journal* [online]. Available at: http://online.wsj.com/article/BT-CO-20120719-709517.html [accessed 15 June 2012].

Yang, C.T. 2012. China drills into shale gas, targeting huge reserves amid challenges. *National Geographic News* [online]. Available at: http://news.nationalgeographic.com/news/energy/2012/08/120808-china-shale-gas/ [accessed 20 June 2012].

Chapter 9
Russian Energy:
Summing up and Looking Ahead

Jakub M. Godzimirski

Introduction

The energy sector plays a special role in Russia. Energy resources secure relative stability on the domestic front by providing legitimacy to a semi-democratic, or semi-authoritarian, regime that can buy the support of the public by using oil and gas windfalls to finance many social and rearmament programmes in Russia. The same energy resources also provide international political clout, making Russia an important player in the global energy game, where the country can use its energy resources to promote its strategic interests. Those are at least two widely held claims about Russia and the role of energy resources in its domestic and foreign policy. This book has sought to offer a deeper understanding of how the recent economic crisis and new global developments have forced the Russian political elite to revise some of their ideas on the importance and usefulness of Russian energy resources as a tool for realizing the country's long-term strategy. The authors of the chapters have presented a range of perspectives on the Russian energy sector, to enable a more nuanced picture of the challenges faced by the Kremlin's energy policymakers in the post-crisis period.

Important here are the country's energy elite – its composition, its ideas on energy and how these ideas have been translated into political action (see Godzimirski, Chapter 1). In order to understand the material base of Russian energy policy, Moe and Kryukov (Chapter 2) enquired into the size and classification of energy resources, first and foremost oil. In Chapter 3, Tatiana Mitrova analysed how the crisis has impacted on the process of formulating the country's energy strategy in a long-term perspective, until 2030 and even beyond. Other contributors have explored the most important actions taken by policymakers and other energy-sector actors in order to make the sector better prepared to meet the future challenges (Baev, Chapter 6); and the impact of Russia's energy resources on the country's foreign policy (Busygina and Filippov, Chapter 5), and on its modernization plans (Averre, Chapter 4). Two chapters dealt with the political and economic aspects of the development of Russia's gas sector: Lunden and Fjærtoft (Chapter 7) on the potential impact on Russian gas exports of introducing a market price for gas; and Laaksonen, Mäkinen and Liuhto (Chapter 8) on the future of the Russian strategic gas projects Yamal and Shtokman, presenting various assessments of future demand for gas in Europe and on the Asian and Chinese market, in view not least of the shale gas revolution.

This final chapter sums up the findings and then explores whether they can support the idea that Putin's regime – the group of people in charge of Russia since late 1999 – has been pursuing a coherent energy strategy, and whether this can be seen as part of an applied grand strategy. We also ask whether attempts at adapting to the new post-crisis situation have been successful, as well as how the future of Russian energy sector is viewed in Russia and abroad, and what are perceived as the main structural constraints on the realization of Russia's energy plans. In other words, this chapter offers a summing up of the findings of the project, as well as some hints as to the possible future of Russian energy policy.

Findings and Challenges

Chapter 1 to 3 aimed at drawing a broader picture of the situation in the Russian energy sector in the post-crisis period, focusing on three key points – *who, what,* and *how.* Godzimirski painted a collective portrait of Russian energy elite, clarifying the formal and informal, institutionalized and non-institutionalized, relationships shaping Russian energy policy under Putin. He also analysed the ideas of a narrow but influential group of policymakers – Putin, Medvedev and Sechin – on the situation in the Russian energy sector and its post-crisis challenges. His analysis shows that Russian leaders have been able to identify the major challenges and have also taken some steps intended to improve its chances of coping. The meetings they organized to discuss ways of dealing with these challenges were highly publicized, officially presented as opening up a new phase of development in Russia's energy sector. However, the effects of their policies have been, to say the least, rather mixed.

On the one hand, several strategic energy projects have been successfully realized, like the Nord Stream pipeline and the opening up of the Bovanenkovo gas field on the Yamal Peninsula. Rosneft and Novatek have managed to close some important deals on close cooperation with key Western companies that seem willing to share technological know-how and economic and financial risks with their Russian counterparts in return for access to Russian resources (Overland et al., 2013). The new BP–Rosneft agreement signed in October 2012[1] has already changed the situation in the Russian energy sector by increasing the state's share in oil production and 'punishing' a powerful group of Russian oligarchs who only one year earlier had dared to challenge Sechin and Putin (Kozlovskiy, 2012). On the other hand, there have also been many delays or even shelving of some projects that were deemed strategic, as in the case of the Shtokman gas field, or the development of closer energy cooperation with China.

---

1   http://lenta.ru/news/2012/10/22/deal/.

What is even worse is that after some 12 years at the top of Russian power pyramid, the re-elected Putin has still to deal with many structural problems he himself identified as crucial in his own 1999 manifesto. This may mean that during the many years he has been responsible for ruling the country he has not been able to cope with those issues – indeed, the situation in some areas, like corruption, has deteriorated even further. What does that tell us about the efficiency of political management of the country in general, and of its energy sector in particular?

In order to answer this specific question we could turn to the chapter by Moe and Kryukov, on Russia's oil resources. They conclude by noting several alarming tendencies in that sector, such as the decreasing size of new discoveries, falling recovery rates and uncertainties about the volume and the quality of the reserves to be developed. Those challenges have been recognized by the country's ruling elite: however, that elite remains basically interested in securing state revenues and has been reluctant to introduce a new system of taxation that will be required if the framework conditions for operations in Russia's energy sector are to improve. Another issue that needs to be addressed is the structure of the sector, which is dominated by a relatively small number of relatively big companies that are not flexible enough to cope with the current resource- and market-related challenges. The fact that Putin and his team have not been able to deal with those issues should be interpreted as a strategic failure that proves their inability to act competently. As that same team is probably set to stay in charge of Russia's energy sector for at least the next six years, we should not expect great improvements in the management of the country's energy resources. It is, however, possible that the involvement of external actors in its development may help to promote better management of the strategic energy projects. Other factors that may spur changes are the competition from other energy producers, and demands coming from actual and potential buyers of Russian energy commodities.

As chapters 7 and 8 have clearly shown, the situation is even more dramatic in the Russian gas sector. While the prices of petroleum products have been more or less liberalized, the price of gas is still largely regulated by state organs, and by politically motivated decisions taken by top members of political elite unwilling to have to face the wave of protests likely to be triggered by a possible gas price hike. In addition, even more than oil sector, the gas sector is dominated by one actor – state-owned Gazprom, which has to cope with serious adaptation problems and seems to be losing ground to domestic and foreign competitors alike (Khaitun, 2012b). This issue is discussed in detail by Lunden and Fjærtoft, and taken up also by Laaksonen, Mäkinen and Liuhto. After examining the domestic and international framework conditions, the latter three conclude that 'Russian gas production and EU demand for natural gas are tightly interlinked'; further, that the future development of Russia's gas sector will depend on at least four factors: production start-up from new fields, the development of unconventional gas resources in Europe and elsewhere, the potential of other gas suppliers, and the policy towards foreign investors whose know-how – and funding – are needed if Russia is to develop new fields in the Arctic.

Dependence on European market, especially as regards gas, is among the structural factors that shape Russian energy policy as well as being one of the main challenges facing the country's policymakers. In the gas market, the situation can be characterized as one of interdependence between Russia and Europe (Casier, 2011; Dickel, 2009). Due to the lack of alternative transport routes and infrastructure needed for shipping gas to other potential markets, Russia's surplus gas production has to be exported to Europe. The EU, in turn, cannot ignore Russia and will need supplies of both gas and oil from Russia for many years to come, even if the production of unconventional gas – and oil – in some parts of Europe should skyrocket and the share of LNG in the European import increase to 30 per cent (for more on a broader framework of EU–Russian energy cooperation see Brunarska et al., 2011).

The heavy energy interdependence between Russia and the EU has created a set of strange incentives that, paradoxically enough, have had a negative impact on Russia's political relations with the EU, as explained in Chapter 5. Busygina and Filippov conclude that the high oil price has made Russian budget overly dependent on revenues generated by oil and gas trade with the EU at the same time when Russian political elite has needed what the authors label 'virtual conflict' with the EU as a way of inoculating the Russian public against harmful Western influences. And yet, Russian and EU energy interests are so intertwined that neither part can afford to withdraw from their energy cooperation, even if there are problems in bilateral relations in other areas and Russia might be tempted to use its energy commodities as a foreign policy tool, as in its relations with Ukraine and some other post-Soviet states. The case of Russia–EU energy cooperation shows how important energy commodities are not only as a source of state revenue, but also in shaping relations with the outside world. It also reveals that there are clear limits to how far Russia can push its energy as a political tool without causing harm to its own interests (see also Goldthau, 2008a).

The lack of a real market alternative is something Russian policymakers are aware of and have been seeking to address, as Baev and Mitrova show in chapters 6 and 3, respectively. Much of what has been happening in Russia's energy relations abroad has been prompted by the desire to become less dependent on Europe and on transit countries. However, as Baev notes in his concluding remarks, 'the various diversification projects have been advancing in uncoordinated surges driven by political expediency, bad personal judgement and corrupt incentives' and the political and economic costs of their implementation may prove prohibitive.

Also Mitrova presents a rather grim picture: for the first time in its recent history, Russia is facing a situation when its ability to export energy commodities is limited by weak demand on foreign markets and not by other factors, like lack of proper investments or sufficient production capacity additions. With the share of revenues generated by production and sales of energy commodities in the state budget mushrooming from 12.2 per cent in 1999 to 52.1 per cent in the first six months of 2012 (Krutikhin, 2012) this situation is recognized as alarming, not only by Russian and Western independent experts but also by Russian politicians (Zonn, 2011).

This issue was hotly debated during discussion on the state budget for 2013 and perspectives until 2015 that took place on 20 September 2012. In the opinion of Minister of Economic Development, A. Belousov, and Minister of Finances, A. Siluanov, the global oil price is the most important indicator Russian policymakers have to watch in a situation when the country's state budget needs an oil price far above 100 USD per barrel in order to remain balanced.[2] President Putin seems very worried about that as well – in his speech at the meeting with the Russian government on 18 September 2012 he seemed pleased that the state budget was almost balanced, but underlined that a situation when the non-oil and gas deficit reaches more than 10 per cent of GDP is unbearable because this makes Russia too exposed to 'the negative impact of the current fluctuations'.[3]

Russia's heavy dependence on revenues generated by sales of oil and gas is something many key Russian policymakers may dislike – but they can do relatively little about it, especially since they have to take into consideration the political costs and risks that the necessary political, social and economic reforms could entail. The easiest way to make Russia's state budget less dependent on oil and gas revenues, and thus make Russia less vulnerable to oil price volatility, is to reduce budget spending, to strengthen budgetary discipline and to end, or at least limit, energy subsidies in Russia. However, the relatively high oil price encourages rather than discourages higher spending in Russia's petroleum-lubricated economy, not least when this spending seems to boost support for the current regime and gives it some legitimacy (Makarkin, 2011). This approach suggests that Russian political elite treats energy resources as an important strategic asset and a means of achieving strategic objectives in other fields (see also Hill, 2004; Goldman, 2008; Shadrina, 2010; Ghaleb, 2011; Tsygankov, 2011). The next part of this concluding chapter will therefore focus on those strategic aspects of energy in the Russian context.

Energy and Russian Grand Strategy

The key focus of this book has been the situation in the Russian energy sector in the aftermath of the 2009 crisis and the process of this sector's adaptation to the new environment. In the title of the book we also asked whether Russia is set to become a hydrocarbons superpower. To answer that question, we must begin by defining how we understand the concept of 'hydrocarbons superpower', and then see whether that is a suitable description of today's Russia. Our understanding of the concept builds on Buzan and Wæver (2003: 24–38), who define a superpower as a country with first- class military and political capabilities and economies to support them, and with global military and political reach, that sees itself and is accepted by others as having this status. In the opinion of Buzan and Wæver,

---

2   http://www.government.ru/stens/20823/.

3   http://eng.news.kremlin.ru/news/4418.

Russia does not meet these criteria and should be rather defined as a 'great power', or a country that does not need to have big capabilities in all sectors and does not need to be present in the securitization processes of all areas in the international system, but must be included in the calculations of other major powers as if it had the economic, military and political potential to bid for superpower status in the short or medium term.

Although Russia does not seem to meet all the criteria for recognition as a superpower in the traditional understanding of the term, it certainly has most of the attributes required of a *hydrocarbons* superpower: a country with first-class energy capabilities and with global energy reach, that sees itself as an energy superpower and is treated as such by others, who have to take Russia into account in their strategic energy calculations. Even more important is the fact that Russia's political elite seems tempted to use energy resources as a weapon and as an important strategic asset in the realization of what we might term *Russia's applied grand strategy*. However, the very same elite is becoming more and more aware of the risks this one-dimensional energy superpowerness may entail – not least since most members of this elite still have fresh memories of the fate of the Soviet Union, which used to be a one-dimensional military superpower; and some of them describe the collapse of the Soviet Union as the greatest geopolitical catastrophe of the twentieth century.

What is then the role of energy in Russian grand strategy? First, let us see how the concept itself has evolved. The classical definition of 'grand strategy' focuses on military aspects and state security (Kennedy, 1991). In his study on the sources of strategic thinking, Barry Posen defined grand strategy as a political-military ends-means chain, a state's theory about how it can best 'cause security' for itself (Posen, 1984:13). In his recent study of threat assessment, Steven Lobell (2009) presents a much broader understanding. He sees grand strategy as not only military, but also fiscal and political in nature; it does not cease at the end of a war or start at the beginning of a war, but is about balancing ends and means in peacetime as well as wartime, involving long-term planning over decades, perhaps longer (2009: 61). Peter Feaver (2009) defines grand strategy as 'the collection of plans and policies that comprise the state's deliberate effort to harness political, military, diplomatic, and economic tools together to advance that state's national interest'.

Grand strategy could therefore be seen as the way in which states employ all major instruments of national policy of importance for achieving the goals defined as 'strategic' by the national ruling power elite. Or it could be described, as John Lewis Gaddis (2009) did in a speech on grand strategy, as the art of relating all of the means at your disposal to the ends you have in view.

According to Clifford Gaddy (2011), energy resources provide Russia with its most important competitive advantage; the main problem facing Russia today is not its endowment with natural resources but the way the rents generated by these resources are used to support inefficient sectors of Russian economy and irrational spending. Prior to the economic crisis, Russia was pleased to be described as an energy superpower (Hill, 2004; IISS, 2006: 177–87; Goldman,

2008; Latynina, 2010) and did not hesitate to use its energy power to promote its interests and to influence the decisions of other actors – although this energy toolbox had already begun to show its clear limitations (Goldthau, 2008a; Orttung and Overland, 2011). It seems safe to assume that the nation's energy resources are treated as an important asset in the realization of Russia's grand strategy under Putin and Medvedev.

This is at least the view of Andrei P. Tsygankov (2011), who in a recent article on Russian grand strategy classifies energy as an important element of the Russian foreign policy toolkit. He notes that energy is treated as a tool of domestic modernization and international influence and presents data supporting this claim. Energy is listed as a strategic tool along with Russia's military, Russia's diplomacy and membership in international organizations, the country's cultural and historical capital and its technological expertise (Tsygankov, 2011).

Tsygankov also explores what have been the results of using energy resources in the pursuit of strategic goals. He notes that Russia plays a key role in the world's energy market, that Russia continues to sell expertise in energy infrastructure abroad, and that by 'capitalizing on high oil prices, it [Russia] strengthened its presence in neighboring economies'; further, that Russia's ties with Europe have been solidified via the construction of strategic pipelines and the joint development of Arctic oilfields, that Russia has completed construction of one oil pipeline to China and has plans to construct two gas pipelines linking Russia with China, that energy cooperation plays an increasingly important part in Russia's cooperation with Turkey, that without its energy capabilities Russia would not have been taken equally seriously by Western and non-Western powers, and, finally, that much of Russia's economic recovery has been due to high oil prices.

This overall positive assessment of the role of energy role in Russian grand strategy notwithstanding, Tsygankov sees also some negative sides of the use of energy as a strategic tool. He states that 'energy disputes undermined its [Russia's] image of a status quo power and reliable energy supplier' and that high oil prices 'slowed down the government's work to reduce reliance on energy exports'. Finally he concludes that 'Although energy remains the country's important comparative advantage, a realistic outlook requires that the Russia is more aggressive in investing in non-energy areas and human infrastructure' (Tsygankov, 2011: 35). This last comment should be read in the context of Tsygankov's assessment of the key strategic challenges Russia will have to cope with in years to come. He cites the volatility of world oil prices as one of those strategic long-term challenges, adding Western expansion in the former Soviet region and the rise of China to that shortlist.

In the case of Russia under Putin and Medvedev, the country's energy resources have been indeed defined as an important economic and political means available to the state for achieving its strategic goals (Gurvich, 2010). While its nuclear weapons are intended to provide national security, Russia's energy commodities have been variously employed in the ongoing international game of interests that could be seen as the implementation of Russia's applied grand strategy (for more on the concept of applied grand strategy, see Murdock and Kallmyer, 2011). At stake when Russian leaders take strategic decisions that will influence the development of the country's energy sector is therefore not only the future of that sector, but to some extent the future of Russia itself. Russia has been hit hard by the recent economic crisis, but has been showing signs of economic recovery. However, the pace of that continued recovery will depend on several factors – on the global oil price, but also on how the Russian decisionmakers decide to tackle the issues haunting the country's economy and not least its energy sector (Abramova et al., 2011; Shmal, 2011). Although Russia is an important global energy player, its policymakers can do relatively little to influence global oil prices. What they can do is to design and implement policies that would result in a more rational approach to their country's energy resources. Here the newly re-elected president has an important part to play.

Putin's Strategic Energy Vision 2012

Putin's relatively swift victory in the presidential elections of 4 March 2012 was a confirmation of his dominant position. It remains to be seen how Putin will use his power and his renewed legitimacy to shape the country's energy policy in his one or two new terms in office. In any case, it is pertinent to see how energy-related issues were framed in his 2012 electoral programme and how he has approached those issues after re-election. By looking at those questions in this final chapter we will focus on what could be termed Putin's strategic vision and his strategic decisions concerning the situation in Russia's energy sector.

Putin's programme was presented to Russian voters on his official campaign site,[4] and elaborated in further detail in seven programme articles published in various Russian media outlets in January and February 2012 (Putin, 2012a, b, c, d, e, f, g).

On his website, energy-related questions were addressed in the fourth part of the campaign programme, on economic issues. Putin admitted that Russia's most serious challenge had been the poor efficiency of its economic system, especially low productivity and low energy efficiency. He pledged to create conditions favourable to the production of energy-efficient products in Russia and the promotion of energy-saving technologies. As analysed in the first chapter of this book, these statements echo his earlier pronouncements on those same issues, made in both his 1999 manifesto and in several statements during his previous presidency. This could also be interpreted as indirect acknowledgment of the failure of his own policies over the dozen-odd years.

Putin elaborated further on those – and many other – issues in a series of programme articles published prior to the March 2012 elections. He acknowledged the link between oil-sector revenues and the wellbeing of the Russian people, and the importance of redistributing oil wealth, but also underlined that Russia's commodity-based economy had no strategic future (Putin, 2012a).

---

4   http://www.putin2012.ru/program.

In his article on the economy (Putin, 2012c), he underscored the importance of the country's natural resources in securing its place on the global economic map, noting that they account for over 25 per cent of Russia's GDP. Putin also described the economy as dependent on global trends, which had made it excessively dependent on the import of consumer goods and on fluctuations in the prices of its main export commodities. He stressed that this situation in which Russia cannot control important developments that have a major bearing on the country was unacceptable, and called for the development of a new economy based on modern technology.

Another issue high on Putin's electoral agenda was the underdeveloped national infrastructure, including gas and electricity networks, and the heavy dependence of the state budget on revenues generated by sales and export of oil and gas. Underlining the need for a balanced budget, he referred to the financial problems faced by the EU and the USA that, in his view, were caused by lack of budget discipline.

In his article on social justice (Putin, 2012e), he warned that the government cannot redistribute more oil and gas revenues to deal with social issues, and that new economic sectors must be developed. In another article (2012c) Putin paid considerable attention to the question of the international division of labour, stressing that Russia should strive to be not only a supplier of commodities and energy but also have cutting-edge technology in at least some sectors. Citing the risk of resource depletion, he called for the development of a new economy, characterized by high labour productivity and low energy intensity.

In the same article (Putin, 2012c), he remarked on Russia's strong competitive advantages, describing the country's vast territory and natural resources – gas, oil, forests, agricultural lands and freshwater resources – as its most important strategic assets. Putin also took up one of the key issues discussed in Russia today – the role of the state in the commodities sector. Here he held that it was a correct decision to increase the influence of the state by curbing the power of Russia's oligarchs and preventing them from selling abroad what he described as 'key assets'. In his view, continued private ownership of those strategic resources would eventually result in external forces taking control over vital elements of the Russian economy.

In his article on national security and Russia's place in the world, Putin (2012f) discussed why Russia needs a strong army, linking this directly to the need to protect the country's natural resources. The world is, he wrote, a place full of unpredictable risks and upheavals, and one in which a zero–sum game is being played. He accused non-identified actors of proposing to lift national control over what he called 'resources of global significance', and warned that Russia would never allow that to happen. Putin also cited this potential threat of Russian sovereignty being challenged by those unnamed actors as a reason for maintaining a strong army.

The incoming president paid even greater attention to energy-related issues in his article on Russia's place in the changing world (Putin, 2012g). Here he underlined the importance of cooperation with the countries of Asia, Latin America and Africa, and defined energy as a key area of such cooperation. He also mentioned Russia's cooperation with the European Union, even proposing the creation of 'a harmonious community of economies from Lisbon to Vladivostok'. Further, he held that more extensive cooperation in the energy sphere should be considered, including the formation of a common European energy complex, and referred to two Russian pipeline projects – Nord Stream and South Stream – as important steps in that direction. He promised reliable and flexible gas supplies to Europe once those projects were realized, stating that construction of those pipelines would put an end to Russia and Europe dependence 'on the political whims of any nation'. Such a solution would, in his view, be especially important for strengthening European energy security in a situation when some states had decided to reduce the role of nuclear energy in their energy mixes. Further, Putin used this opportunity to fire a salvo against the EU's Third Energy Package, which he described as aimed 'at squeezing out integrated Russian companies'. Here he openly accused this legal move of being damaging to relations between Russia and the EU. He then went on to accuse the West of not opening its markets to Russian companies, which are often 'brushed aside' when they want to invest – whereas, he noted, Russia has been 'opening up the most attractive areas of our economy to foreign investors, granting them access to the 'juiciest morsels', in particular, our fuel and energy complex'. Putin also mentioned Russia's role as a global supplier of safe and modern nuclear technology, which should make the country an attractive cooperation partner.

After taking office, the new president has taken several steps that show he is aware of the challenges faced by Russia and its energy sector. In May 2012, only one day after his official inauguration, Putin issued several executive orders, including one on the future economic policy and another one on Russia's foreign relations.[5] In the first, he ordered the Russian government to draft a federal law setting out a mechanism for using federal budget oil and gas revenues and for forming, using, and managing the funds in the Reserve Fund and the National Welfare Fund.[6] In the second one, he ordered the Ministry of Foreign Affairs 'to develop a mutually beneficial energy partnership with the aim of creating a common European energy sector, ensuring strict compliance with existing bilateral and multilateral contractual obligations'.[7]

On 15 June 2012 Putin decided to establish a new Presidential Commission for Strategic Development of the Fuel and Energy Sector and Environmental Security with Igor Sechin as its secretary, and the president himself as its head. The task of that Commission would be 'to coordinate the work of the federal and regional executive authorities, local self-government authorities, and organisations in developing the fuel and energy sector, ensuring industrial, energy and environmental security, and rational use and effective management of the minerals and raw materials resource base'.[8] By the end of November

---

5   For an overview of those orders see: http://eng.kremlin.ru/acts?date=May+7%2C+2012.
6   http://eng.kremlin.ru/acts/3758.
7   http://eng.kremlin.ru/acts/3764.
8   http://eng.state.kremlin.ru/commission/29/news/4023.

2012 the Presidential Commission had held two meetings – on 10 July and then on 23 October 2012. At the first meeting Putin presented his own vision of the Commission's responsibilities: it had been created, he stated, to coordinate joint efforts in Russia's energy sector, involving participants of the energy market, scientists, representatives of public organizations and law enforcement agencies who were to analyse strategic issues with the direct participation of the government, which was responsible for adopting the final decisions on various matters.[9]

At its first meeting the Commission focused on such issues as the expansion of the geographical scope of hydrocarbons production in the near future, the development of a high-tech energy industry, the expansion of international cooperation in the energy field, the role of the state in the energy sector and management of state-owned companies, the need for greater transparency in the sector, and, last but not least, environmental issues.[10]

The second meeting of the Commission focused on three key matters – the draft Doctrine of Russia's Energy Security, the long-term prospects of strengthening Russia's position in the global energy markets, and electricity-sector cooperation with the EU, Russia's closest partner and major consumer.[11]

The fact that Putin has paid so much attention to questions related to the development of the country's energy sector in his electoral programme, has been able to identify key challenges, understand the impact that developments in that sector will have on other policy areas and has decided to set up and head a new presidential commission for coordinating its development – all this demonstrates that he has adopted a strategic approach to energy resources and recognizes their importance in Russian grand strategy. For instance, it is obvious that realization of an ambitious rearmament programme Putin proposed in one of his programme articles – issuing an executive order on its implementation the day after his inauguration[12] – will not be possible without huge revenues generated by the country's energy sector.

However, the fact that Putin recognizes the strategic importance of Russia's energy resources does not mean that he will succeed in realizing all the goals of Russia's grand strategy, the most important of which, according to Tsygankov (2011), are to become 'an independent center of power and influence' and to preserve dominance in the former Soviet region. For that to happen he would have to adopt not only a strategic but also a more comprehensive approach to the various problems facing his country.

Will Putin Succeed?

Western and Russian observers alike have commented on Putin's visions. One of them (Silayev, 2012), wrote that while four years ago Putin's calls for making

---

9   http://eng.state.kremlin.ru/commission/29/news/4150.

10 http://eng.state.kremlin.ru/commission/29/news/4150.

11 http://eng.state.kremlin.ru/commission/29/news/4553.

12 http://eng.kremlin.ru/acts/3777.

Russia less dependent on selling raw materials had sounded like a wish, by the beginning of 2012 they were sounding more like an imperative. The key reason is that a resource-based economy cannot secure the rapid economic growth necessary to meet the growing expectations of Russians, and that the global markets have become far more turbulent, making predictions on future oil prices almost impossible. As argued in Makarkin (2011), making the Russian economy less dependent on raw materials is not only an economic necessity but also a political one, as the legitimacy of the current regime hinges on its ability to deliver stability and better perspectives.

The choices made today by key Russian policymakers will have major consequences for the further development of the country. The challenge is that they must devise and implement a balanced strategy for the development of the energy sector, currently the cornerstone of the national economy. This strategy will have to ensure the sustainable and balanced development of the entire Russian economy.

This dilemma has been recognized by many observers. As Clifford Gaddy (2011) put it in analysing the role of Russia's energy sector, 'any dream of creating a "non-oil" Russia that could perform as well as today's commodity-based economy is unrealistic', not least due to the fact that oil is, in his view, Russia's comparative advantage, and the energy sector as such is the most competitive part of its economy (on Russia's rather poor competiveness, see Drzeniek and Prazdnichnykh, 2011). In Gaddy's view, the problem is not the abundance of oil and gas in Russia, but the way the oil rent is used to support other ineffective parts of the economy.

In the final chapter of his insightful study on 'The battle for oil and power in Russia', Thane Gustafson (2012: 488–501) considers three visions of the Russian future: Medvedev's high-tech modernization, Kudrin's back-to-basics vision, and Putin's 'trust the state and stay the course' approach. Now, with Kudrin gone, Medvedev clearly marginalized and Putin firmly in charge, it may seem that it is Putin's vision that will serve as guidance for policymaking for many years to come. What could this mean for the future of Russia as an energy nation? The key question is not *whether* Russia will remain an important energy player, since the energy resources will secure it a central place in the international energy game (on the resource base, see Ministry of Natural Resources 2012), but rather *what kind of* energy player Putin's Russia will be in years to come, and what will be the results of the continuation of Putin's policy in the energy sector. Here there are several possible energy futures for Russia – not all of them bright.

Aleksey Moyseyev (2011) is straightforward in writing that the dilemma facing Russia is not whether the country ought to develop as an energy-based economy, but rather what sort of an energy-based economy Russia should be. In his view, the choice facing Russia is between becoming a 'developed' supplier of resources to the world market, or a 'developing country'.

The first path is the one followed by countries like Canada, Norway and Australia; the second has been the fate of Nigeria, Indonesia and the oil-exporting countries of Latin America. The trajectory of the first group shows, writes Moyseyev, that 'it is possible to achieve sustainable economic development and become a world leader in providing high standard of living, even if your economy is based on such a seemingly low-tech sector as the extraction of mineral raw materials'.

Another leading Russian economist is Vladimir Mau – since 2002 the rector of the Russian Presidential Academy of National Economy and Public Administration, as well as the head of the Strategy 2020 group, and since 2011 a member of Gazprom's board of directors. He notes that Putin's Russia may follow one of three paths of development (Mau 2011): the inertia scenario, the growing raw materials exports scenario, and third, the innovation scenario. Summing up what each of those scenarios could mean for Russia, he writes: 'these scenarios might be described as the Nigerian, Mexican and Australian scenarios' (2011: 23). The Nigerian scenario is the situation of a country with huge fuel and energy resources that has seen stagnation and degradation of its political and economic institutions and a series of political and economic crises. The Mexican scenario describes a country that has received massive windfalls from sales of oil, resulting in significant but uneven economic growth. And finally, the Australian scenario can be described as the success story of a country that is rich in natural resources and has used them to diversify its economy and improve its international standing (Mau 2011: 23).

Which of these three developmental paths will Russia follow in the twenty-first century? Much hinges on how the ideas on the role of energy held by those in charge today will influence policy choices in the years to come. Although it seems that both Putin and Medvedev realize that Russia is today at an 'energy crossroads', it remains to be seen whether they will have the courage and determination to do what must be done to push Russia in the right direction.

However, whether Russia will be able to enter and follow the right path will depend not only on the will and choices of the country's leaders but, perhaps even to a greater degree, on their ability to cope with the various structural factors that constrain their policy choices and the effectiveness of their actions.

Some of these factors, like the market-driven processes of oil-price formation, have an obvious impact on the development of the energy sector and the Russian economy. With other factors, like the US-driven shale gas and oil revolution, the impacts are as yet far less clear, although there is growing realization in Russia that this may be a real game changer, or, as President Putin put it, may completely reshape the market[13] (see also Mitrova, 2012; Grätz, 2012; Maugeri, 2012). The structural dependence on the European gas market and on gas transit countries seems to have been recognized and partly tackled by Russia's decisionmakers, as Baev shows in Chapter 6 of this book, where he describes some partly chaotic but successful attempts at diversifying routes and markets.

Discussing the impact of structural factors on Russian energy policymaking involves a kind of Catch-22 situation. On the one hand, energy policy must be

---

13   http://ria.ru/economy/20120411/623107721.html.

viewed as an important part of Russia's economic policy. On the other hand, Russian energy policy is an important factor that shapes how the Russian economy in general can operate, and thus should be seen not only as a part of the economy, but also as one of factors that impact on the entire economic and even the political and social development of the country (Gaddy and Ickes, 2010; 2011).

Richard Connolly (2011) cites the overdependence on natural resources as an important negative structural characteristic of the Russian economy that remains unsolved and impedes modernization. Although in formal analytical terms this overdependence should not be treated as a structural factor, because Russian policymakers *could* eliminate it by implementing various measures, it must be recognized as an important element of the framework conditions affecting the whole Russian economy.

Connolly (2011) presents his own list of factors that influence economic developments in today's Russia, based on a thorough analysis of the World Economic Forum's competitiveness indexes from 2004 to 2010. His list includes corruption, access to financing, tax regulations, crime and theft, inflation, inefficient bureaucracy, tax rates, inadequately educated workforce, poor work ethic, inadequate infrastructure, policy instability, foreign currency regulation, government instability, and restrictive labour regulations.

Although not all these factors have a direct bearing on the development of the energy sector, they shape the environment in which strategic decisions are made and policies are implemented. Russia's poor performance during the global financial crisis highlighted a number of problems. Its dependency on the revenues generated by production and export of natural resources became evident. Natural resources accounted for around 80 per cent of Russia's exports, around half of public sector revenues, and around a quarter of GDP. Fluctuations in international commodity prices did indeed shape economic developments in Russia (Connolly, 2011: 431).

Zheleznova and Panfilenko (2011) cite another three key factors shaping Russia's energy policy today: the shortage of developed reserves of oil and natural gas that can allow the country to meet its international commitments and rising domestic demand; the growth in prices for energy resulting from liberalization of the economy and getting them to reach international levels; and the fact that energy resources are concentrated east of the Urals, whereas the main areas of energy consumption are in the Urals and the European part of Russia, which increases the cost of energy supplies.

In addition to being overly dependent on the EU market, Russia faces several other structural geographical challenges. Khaitun (2012a) notes the lack of ice-free ports, the limited number of available export routes, and Russia's inability to control the energy assets of neighbouring countries and their energy export routes.

Although Russia has huge deposits of oil, gas and coal that should, in theory, help it to meet the supply challenge, there are several factors that may negatively influence developments in this sector and entail delivery problems (Riley and Umbach, 2007). In analysing Russian energy security, Khaitun (2012a) presents a mixed picture of the perspectives. On the one hand, today the energy sector delivers what is needed and also has plans for meeting future challenges. On the other hand, the development of that sector has been neglected, and is influenced by many negative factors that affect the development of the overall Russian economy. Connolly (2011) describes the lack of proper domestic and foreign investments in Russia as perhaps the most important factor currently impacting on the country's ability to modernize and diversify its economy, including the energy sector.

In the mineral extraction sector, investments will need to reach the level of 30 per cent of GDP in order to replace the obsolete assets with the new ones. However, in 2010 investments in Russia reached a level of only 20 per cent of GDP – an investment gap that will have dire consequences for the future development of the economy, because 46 per cent of fixed assets in that sector are today depreciated (Zaslavskaya, 2011; Chichkin, 2011).

As the ability of Russian energy companies and the Russian state to secure the necessary level of investments in the energy sector depends on access to proper funding, Khaitun (2012a) adds market-related threats to the list of the challenges facing the country. He warns that the smaller volumes of oil and gas exports combined with falling prices for those commodities could undermine Russia's economic stability by reducing its currency earnings and its budgetary revenues.

At the same time Russia has become more dependent on the energy rent than earlier. Whereas in 2004 the country's state budget needed an oil price of USD 20 per barrel in order to be balanced (Gavrilenkov and Stroutchnevski, 2010), balancing the 2012 budget required a price of more than USD 117 (Khaitun 2012a).

In addition to those purely economic factors, historical and social factors will also play a part. These include the country's human capital, its political culture and Russia's path dependence, i.e. the way historical experiences influence the political decisions of today. Hedlund (2005) has discussed in detail the question of Russian path dependence that results *inter alia* in what has since been termed the Sovietization of Russian politics (Kryshatnovskaya and White, 2009). The issue of human capital was discussed in Chapter 1 of this book, especially as regards those in charge of the country's energy sector.

Most of the other chapters of this book have investigated and explained how the framework conditions in which Russian energy policy is being designed and implemented are shaped by how the Russian 'dual state' (Sakwa, 2010) – with its specific political culture and Putin with his specific vision of Russia and energy (Putin, 2000) at the helm – interacts with the global energy market. It remains to be seen how the future of Russia and its energy sector will be shaped by Putin's vision and how the realization of that vision may be constrained by a range of structural factors, as described above. As the contributions to this volume have shown, Russia's energy resources are indeed its strategic asset and may provide many opportunities – but how they are used and the effect of that use will depend on those in charge, and on how their visions are translated into political action.

References

Abramova, E., A. Apokin, D. Belousov and K. Mikhaylenko. 2011. *On the development of the Russian economy in 2011 and forecast for 2012–2014*. Available at: http://projects.chass.utoronto.ca/link/meeting/ctryrep/RUS_201110.pdf [Accessed 20 June 2012].

Brunarska, Z., A. Jarosiewicz, A. Łoskot-Strachota and I. Wiśniewska. 2011. Between energy security and energy market integration. *OSW Report* 6. Warsaw: Osrodek Studiów Wschodnich.

Buzan, B. and O. Wæver. 2003. *Regions and Powers: The Structure of International Security*. Cambridge: Cambridge University Press.

Casier, T. 2011. The rise of energy to the top of the EU–Russia agenda: from interdependence to dependence? *Geopolitics*, 16(3), 536–52.

Chichkin, A. 2011. Poiznosilis'. Uroven' iznosa osnovnykh fondov v Rossii namnogo vyshe, chem v drugikh stranakh BRIKS [Worn out. Level of depreciation of main funds in Russia are higher than in other BRIKS countries]. *Rossijskaya gazeta*, 5 July.

Connolly, R. 2011. Financial constraints on the modernization of the Russian economy. *Eurasian Geography and Economics*, 52(3), 428–59.

Dickel, R. 2009. *Energy Interdependence between EU and Russia: Security of Supply / Security of Demand* [Presentation]. Energy Charter Secretariat for CEPS, 29 September 2009. Available from http://www.ceps.eu/files/2Dickel.pdf [Accessed 20 June 2012].

Drzeniek H., M and A. Prazdnichnykh 2011. *The Russia competitiveness report 2011: Laying the foundation for sustainable prosperity*. Davos: World Economic Forum.

Feaver, P. 2009. What is grand strategy and why do we need it? Available at: http://shadow.foreignpolicy.com/posts/2009/04/08/what_is_grand_strategy_and_why_do_we_need_it [Accessed 20 June 2012].

Gaddis, J.L. 2009. What is grand strategy? Available at: http://www.duke.edu/web/agsp/grandstrategypaper.pdf [Accessed 20 June 2012].

Gaddy, C. and B. Ickes. 2010. Russia after the global financial crisis. *Eurasian Geography and Economics* 51(3): 281–311.

Gaddy, C. and B.W. Ickes. 2011. The Russian economy through 2020: the challenge of managing rent addiction. In *Russia in 2020. Scenarios for the Future*, edited by Maria Lipman and Nikolay Petrov. Washington DC: Carnegie Endowment for International Peace, 165–86.

Gaddy, C. 2011. Will the Russian economy rid itself of its dependence on oil? Available at http://valdaiclub.com/economy/26960.html [Accessed 20 June 2012].

Gavrilenkov, E. and A. Stroutchenevski. 2010. Russia: Poised for growth, but constrained by policy. Presentation, December 2010, Moscow: Troika Dialogue Available at: events.bfm.ru/upload/events/file_manager/22_7_cnhextytvski.ppt [Accessed 20 June 2012].

Ghaleb, A. 2011. *Natural gas as an instrument of Russian state power*. Strategic Studies Institute, U.S. Army War College 2011 [Online]. Available at http://purl.fdlp.gov/GPO/gpo14259 [Accessed 2 December 2012].

Goldman, M.I. 2008. *Petrostate: Putin, Power, and the New Russia*. Oxford: Oxford University Press.

Goldthau, A. 2008a. Resurgent Russia? Rethinking Energy Inc. *Policy Review* [Online] 147:53-63 Available at: http://www.hoover.org/publications/policy-review/article/5714 [Accessed 20 June 2012].

Goldthau, A. 2008b. Rhetoric versus reality: Russian threats to European energy supply. *Energy Policy*, 36(2), 686–92.

Grätz, J. 2012. Unconventional resources: The shifting geographies and geopolitics of energy. In *Strategic Trends 2012. Key Developments in Global Affairs*, edited by Daniel Möckli. Zurich: Center for Security Studies, 79–102.

Gurvich, Y. 2010. Neftegazovaja renta v rossiskoj ekonomike. *Voprosy ekonomiki*, 11: 4–24.

Gustafson, T. 2012. *Wheel of Fortune: The Battle for Oil and Power in Russia*. Cambridge MA: Belknap Press of Harvard University Press.

Hedlund, S. 2005. *Russian Path Dependence: A People with a Troubled History*. London: Routledge.

Hill, F. 2004. *Energy Empire: Oil, Gas and Russia's Revival*. London: The Foreign Policy Centre.

IISS. 2006. Russia: An Emerging Energy Superpower?, in *IISS Strategic Survey 2006*. London: International Institute of Strategic Studies.

Kennedy, P. 1991. Grand strategy in war and peace: toward a broader definition, in P. Kennedy (ed.) *Grand Strategies in War and Peace*. New Haven,CT: Yale University Press.

Khaitun, A. 2012a. Energobezopasnost' pod voprosom. Neraduzhnye perspektivy razvitiya situatsii na toplivnom rynke [Energy safety is under question. Bad situation at a petroleum market]. *Nezavisimaya Gazeta NG-Energia* [Online] http://www.ng.ru/energy/2012-01-17/9_fuel.html [Accessed 20 June 2012].

Khaitun, A. 2012b. 'Gazprom' na glinyanykh nogakh. Rekonstruktsiya gazovoj otrasli v Rossii nedopustimo zatyagivaetsya [Gazprom on clay feet. Reconstruction of the gas sector in Russia is taking too long time]. *Nezavisimaya Gazeta Energia* (9 October at http://www.ng.ru/energy/2012-10-09/9_gazprom.html).

Kozlovskiy, S. 2012. Putin razreshil. Prezident odobril prodazhu poloviny TNK-BP gosudarstvennoj 'Rosnefti' [Putin allowed. President allowed selling half of TNK-BP to state agency Rosneft]. [Online: *lenta.ru*]. Available at: http://lenta.ru/articles/2012/10/11/tnkbp/ [Accessed 20 June 2012].

Krutikhin, M. 2012. Privatizatsiya v neftegazovoj otrasli Rossii: sostoyanie i perspektivy [Privatization in Russian petroleum sector: current state and perspectives] . Presentation at *IMEMO Seminar on 6 November 2012* Moscow: IMEMO [Online] Available at http://www.imemo.ru/ru/conf/2012/06112012/KMI_06112012.pdf [Accessed 3 December 2012].

Kryshtanovskaya, O. and S. White. 2009. The Sovietization of Russian Politics. *Post-Soviet Affairs*, 25(4), 283–309.

Latynina, J. 2010. An energy superpower by default. *The Moscow Times*, 10 March.

Lobell, S.E. 2009. Threat assessment, the state and foreign policy: a neoclassical realist model, in J.W. Taliaferro, S.E. Lobell and N.M. Ripsman (eds), *Neoclassical realism, the state, and foreign policy*. Cambridge: Cambridge University Press.

Makarkin, A. 2011. The Russian Social Contract and Regime Legitimacy. *International Affairs*, 87(6), 1459–75.

Mau, V. 2011. The role of state and creation of a market economy in Russia. *BOFIT Discussion Paper 23*. Helsinki: Bank of Finland's Institute for Economies in Transition BOFIT.

Maugeri, L. 2012. Oil: The Next Revolution. The Unprecedented Upsurge of Oil Production Capacity and What It Means For The World. *Geopolitics of Energy Project*. Cambridge, MA: Harvard Kennedy School, Belfer Center for Science and International Affairs. [Online] Available at: http://belfercenter. ksg.harvard.edu/files/Oil-%20The%20Next%20Revolution.pdf. [Accessed 3 December 2012].

Ministry of Natural Resources. 2012. Gosudarstvennyj doklad 'O sostoyanii i ispol'zovanii mineral'no-syr'evykh resursov Rossijskoj Federatsii v 2010 godu'. Moscow: Ministry of Natural Resources, Russian Federation.

Mitrova, T. 2012. Slantsevaya neft – yest' li u vas Plan B? [Shale oil – do you have a plan B?] *IMEMO seminar on Shale oil 2 October 2012*. Moscow: IMEMO, at http://www.imemo.ru/ru/conf/2012/02102012/MIT_02102012.pdf [Accessed 20 November 2012].

Moyseyev, A. 2011. *Rossiya i mir*. Moscow: Strategy 2020.Available at: http://strategy2020.rian.ru/load/366062692 [Accessed 20 June 2012].

Murdock, C. and K. Kallmyer. 2011. Applied grand strategy: making tough choices in an era of limits and constraint. *Orbis*, 55(4), 541–57.

Orttung, R.W. and I. Overland. 2011. A limited toolbox: Explaining the constraints on Russia's foreign energy policy. *Journal of Eurasian Studies*, 2(1), 74–85.

Overland, I., J.M. Godzimirski, L.P. Lunden and D. Fjærtoft. 2013. Rosneft's offshore partnerships: the re-opening of the Russian petroleum frontier? *Polar Record*, 49(2), 140–53.

Posen, B.R. 1984. *The sources of military doctrine: France, Britain, and Germany between the world wars*. Ithaca, NY: Cornell University Press.

Putin, V. 2000. Mineral'no-syr'evye resursy v strategii razvitiya Rossijskoj ekonomiki, in N.N. Moiseev and S.A. Stepanov (eds), *Rossiya v okruzhayushchem mire: 2000 (Analiticheskiy ezhegodnik)*. Moscow: MNEPU.

Putin, V. 2012a. Russia muscles up – the challenges we must rise to face (published 16 January). Available at: http://premier.gov.ru/eng/events/news/17755/ [Accessed 20 May 2012].

Putin, V. 2012b. Russia: The ethnicity issue (published 23 January). Available at: http://premier.gov.ru/eng/events/news/17831/ [Accessed 20 May 2012].

Putin, V. 2012c. Economic tasks (published 30 January). Available at: http://premier.gov.ru/eng/events/news/17888/ [Accessed 20 May 2012].

Putin, V. 2012d. Democracy and the quality of government (published 6 February). Available at: http://premier.gov.ru/eng/events/news/18006/ [Accessed 20 May 2012].

Putin, V. 2012e. Building justice: A social policy for Russia (published 13 February). Available at: http://premier.gov.ru/eng/events/news/18071/ [Accessed 20 May 2012].

Putin, V. 2012f. Being strong: National security guarantees for Russia (published 20 February). Available at: http://premier.gov.ru/eng/events/news/18185/ [Accessed 20 May 2012].

Putin, V. 2012g. Russia and the changing world (published 27 February). Available at: http://premier.gov.ru/eng/events/news/18252/ [Accessed 20 May 2012].

Riley, A. and F. Umbach. 2007. Out of gas: looming Russian gas deficits demand readjustment of European energy policy. *Internationale Politik* (Spring): 83–90.

Sakwa, R. 2010. *The Crisis of Russian Democracy: The Dual State, Factionalism and the Medvedev Succession*. Cambridge: Cambridge University Press.

Shadrina, E. 2010. *Russia's foreign energy policy: norms, ideas and driving dynamics*. 18/2010, Electronic Publications of Pan-European Institute. Tampere: Pan-European Institute.

Shmal, G. 2011. Rozhdenie novoj ekonomiki Rossii dostizhimo cherez reorganizatsiyu otechestvennogo neftegazovogo kompleksa. *Ekonomika i TEK segodnya* [Online] Available at: http://is.park.ru/doc.jsp?urn=49862293 [Accessed 20 May 2012].

Silayev, N. 2012. Politekonomia pozdnego Putina. *Expert* [Online]. Available at: http://expert.ru/expert/2012/05/politekonomiya-pozdnego-putina/ [Accessed 20 May 2012].

Tsygankov, A.P. 2011. Preserving influence in a changing world. Russia's Grand Strategy. *Problems of Post-Communism*, 58(2), 28–44.

Zaslavskaya, O. 2011. V bank za 'dlinnym' rublem. Pochemu kreditnye uchrezhdeniya ne sposobny obespechit' rossiyskoy modernizatsii neobkhodimyy denezhnyy potok [To the bank for a bigger ruble. Why can't the credit organizations provide necessary means for Russian modernization]. *Rossiyskaya gazeta* [Online]. Available at: http://www.rg.ru/2011/07/05/finansy-modernizaciya.html [Accessed 20 May 2012].

Zheleznov, V. and E. Panfilenko. 2011. Prakticheskie vyzovy dlya Rossii. Problema ekonomii v global'nom i mestnom masshtabe [Practical challenge to Russia. Problems of economics at global and local levels]. *Nezavisimaya Gazeta NG-Energia*, [Online]. Available at http://www.ng.ru/energy/2011-12-13/13_vyzovy.html [Accessed 20 May 2012].

Zonn, I. 2011. Riski neftyanogo rynka. Dlya byudzheta minus dollar za barrel' oznachaet milliardnye poteri [Risks of the petroleum market. Minus dollar for a barrel means losing millions in a budget]. *Nezavisimaya Gazeta NG-Energia*.

Index